Structured
Requirements
Definition

Published by
Ken Orr and Associates, Inc.
Topeka, Kansas

structured
requirements
definition

by Ken Orr

First printing 1981
Second printing 1981
Third printing 1982

Editors:
Karen Howard Brown
Marlene B Orr

Artist:
Bob Senogles

Library of Congress Catalog Card Number: 81-80846

Orr, Kenneth T.
    Structured requirements definition.
Topeka, KS: Orr, Ken, & Associates, Inc.
8104   810226

ISBN 0-9605884-0-X

Printed in the United States of America

. . . to

Nancy, Judy, Katy, Paige
and Marlene . . .

Contents

****

Foreword

```
******************************************************
```

I am very pleased to to be able to introduce Ken Orr's new book, **Structured Requirements Definition**, as I have looked forward to its publication with great interest.

I would like to focus particularly on Orr's statement at the beginning of Chapter 2: "Bad systems are complex, hard to change, hardware and software dependent, and monolithic."

Why is it so important to have systems that are simple and easy to change? Orr gives several reasons: to obtain systems that are smaller, cheaper, and less time-consuming in their development. I agree with him, but

I would like to pursue the question a little further.

Why design systems that are easy to change? To me, the reason, which has never been clearly stated, can be found in the distinction between the human mind and the computer. The content of the human mind is ever changing. The content of the computer is data, and data never changes except when it is updated by programs.

Human Organizations[1] are changing all the time. The structures of systems change if, and only if, someone works to modify them. It is not sufficient to modify the contents of the systems (files, data, and programs), however. It is more important to modify the structures of the systems so we can obtain a good correlation between the Organization and the system of data.

This book will give you a very good introduction to the process of defining requirements for the end-user.

J.-D. Warnier
Paris
September 1980

---

[1] The term "Organization," with a capital O, has been chosen for the French term "entreprise." It should be interpreted in the large sense to cover commercial or noncommercial organizations.

Preface

****

    Books are written for a variety of reasons. But whatever the reason, a book, any book, is the product of a great many people. Such is certainly the case with this one.

    Structured Requirements Definition is the product of those who contributed to the development of the methodology, those who assisted in the refinement of the methodology, and those who contributed to the editing, art, and production of the book.

    Many major concepts behind **Structured Requirements Definition** were the result of pioneering work by a Langston, Kitch and Associates project team that included Pete Kitch, Bill Bryant, Morris Nelson, and Terry Young. It is to this team that we owe the development of the entity diagram, and the working out of the basic concept of the functional flow.

Others at Langston, Kitch who played a large part in refining our techniques on real projects were Bob Langston, Art Larsen, and David Taylor.

I am also indebted to Terry Swanson, Gordon Harding, Charles Finley, David Higgins, Jacki Summerson, Jim Highsmith, and Gary Wilke for their observations concerning the methodology.

On our staff, Bob Senogles labored long and hard to make sense of the multitude of graphs and figures.

Finally, I would like to thank Karen Howard Brown and Marlene Orr for their tireless efforts to make my ramblings intelligible. Moreover, they did this not once, but twice.

This book represents something of a stop-motion photograph of a moving target. The science of systems development will continue to evolve, but from time to time it is necessary to stop and document the current state of the art.

Ken Orr

Introduction/1

****

It must be remembered that there is nothing more
difficult to plan, more doubtful of success,
nor more dangerous to manage, than the
creation of a new system. For the initiator
has the enmity of all who would profit by the
preservation of the old institutions and
merely the lukewarm defenders in those who
would gain by the new ones.

Machiavelli, **The Prince**

ART AND SCIENCE

There is but a thin line between art and science. Indeed, all the sciences as we know them today first began as arts. But just where does the transformation from art to science take place?

Perhaps the line between art and science is crossed when the artisan first becomes concerned with understanding "why" as well as "what." Perhaps it happens when traditional approaches are applied to a slightly larger problem, and failure results. Perhaps it occurs when the invention of a new tool taxes the ability of the artisan in its application.

Whatever the reason, there comes a point in the maturation of a field when it passes from a craft to an engineering discipline and then to a science. In recent years, that change has occurred in the field I will call systems science (informatics).

Those of us actively involved in this area might be too close to it to notice these subtle modifications. But noticed or not, changes in the field have increased in frequency in the last several years. Concomitantly, there has been a rising interest in improving the tools and techniques available to systems scientists.

Unfortunately, there is always a delay between the "leading edge" thinking in any area and communication of these ideas to practitioners in the field. This book is intended to relate the most important knowledge garnered from the expansion of the science of systems building.

SYSTEMS METHODOLOGIES

There is a plethora of books on systems methodologies available today. And all of them recommend adherence to a certain pattern: to get first things first.

These books on systems design universally propose a series of development phases. Some call the phases by one set of names, others use different titles. But the general philosophy is much the same. The lists on page 4 show but a few of the different systems life cycles proposed by various authors over the last twenty years.

What's in a Name

A rose by any other name is still a rose. No matter what creative names are attached to the planning, requirements definition, or design phases, it doesn't change the situation very much. There are a limited number of ways to do things right. Although each of us likes to imagine that we have uncovered the secret of the universe for the first time, that is rarely the case. When intelligent, industrious individuals study the same subject, they often come up with the same conclusions, hence the basic agreement upon the phases in the systems life cycle.

With all this basic agreement, it would appear to be a simple matter for systems scientists to concur about the best method for developing systems. Unfortunately, such accord has proved more difficult to achieve than might be expected. Systems scientists are no quicker to admit errors or to allow that someone else's approach might have some merit than are individuals in any other profession.

Optner (1960)
1. Analyze the present systems
2. Develop a conceptual model
3. Test the model
4. Pilot installation of the new system
5. Full installation of the new system

Ellis (1962)
1. Define mission objective
2. Develop mission design
3. Requirements analysis
4. Systems specifications
5. Capabilities analysis
6. Systems design
7. Fine structured design

Heany (1968)
1. Develop or refine information requirements
2. Develop gross systems concepts
3. Obtain approval
4. Develop detail design
5. Test
6. Implement
7. Document
8. Evaluate

Fisher (1969)
1. Identification of the problem
2. Description of the problem
3. Design of the system to solve the problem
4. Program the system
5. Implement the system
6. Support the system
7. Fine structured design

Measuring Success

Although other sciences have developed experimental means of testing one set of ideas against another, the science of systems development has not, making it extremely difficult to compare one method or theory with another. Since there are no clear measures of success, what can be considered the criteria for a successful system? User satisfaction, efficient utilization of automated equipment, flexibility, or simply a system installed on time?

In many cases, even the obvious measures may be misleading. A successful new system installed on time and under budget may finally prove too expensive to operate or too inflexible to change. A system that fits one user or one set of business conditions may very well become an obstacle to future development. Other professions have had generations to observe their handiwork, but systems science is new, and what appears to be a good idea today may prove to be a poor one later.

But even with all the difficulty of measuring the success of various approaches, there have been some attempts at evaluation. Dijkstra and Constantine, for example, have introduced a means of measuring pieces of systems called programs. Indeed, many of Dijkstra's initial observations regarding program structure had to do with intellectual clarity, and Constantine's with the cost of development and maintenance. Warnier and others have appealed to the concept of logical correctness as a measure. Each of these ideas has value and, in fact, they all appear to be related.

Good systems are robust, i.e., they remain correct over time. Good systems should be flexible (adaptable) but resistant to changes that make them less robust.

This is important since many writers on systems development often ignore the fact that making a system easier to change may also make it easier to disrupt, corrupt, or malfunction.

It is difficult, if not impossible, to predict what an individual or group of individuals will do over time. And since information systems are identified, defined, designed, and operated by human beings, the business of methodically building a robust system is an inherently risky undertaking.

But risk or no, circumstances dictate that we now have to create systems that are more robust than those developed in the past. To provide better products and services in this modern world, organizations need better information; thus, information systems represent the very central nervous system of any organization. Today, it is possible to destroy a large organization simply by disrupting its information system.

Reliability and robustness, then, go hand in hand, and reliability is increasingly important in systems intended to support every aspect of operations. The failure of a computer-based information system to change and still remain correct can severely limit the future of an organization.

Expanding the Discipline

This book presents an extension of the process of logical systems building that was delineated earlier in **Structured Systems Development.**[1] Although in the last

---

[1] Orr, Kenneth T. **Structured Systems Development**. New York: Yourdon, Inc., 1977.

few years the notion of structured systems design has matured and been expanded, the basic approach advocated in **Structured Systems Development** is still valid. As far as systems design is concerned, the outputs are clearly the right place to begin, and that argument will be restated in Chapter 2. It is not always possible, however, to elicit the required outputs directly from the user.

Thus, we have labored to develop a requirements definition process that works from the definition of the problem to the definition of the outputs. But even that is not enough. It is also necessary to develop a planning and definition phase that moves from an initial problem statement (symptoms) to correct definition of the problems and scope (systems identification).

Systems science is extremely new. It is the outgrowth of the pressing need to develop complex, highly interrelated information and control systems. Over the last decade we have seen the convergence of a great many theories and practices. Thus, as one set of ideas has complemented another, we have been able to place the systems development process on firmer foundations.

Good systems are not accidental. Indeed, like good methodologies, good systems all look very similar; it is only the bad ones that show much variety. Unfortunately, as in medicine, systems science has been mainly concerned with the study of pathologies, things that don't work. Only recently have systems scientists begun to investigate what works and why it works.

Background/2

****

...the task of natural science (is) to show
that the wonderful is not incomprehensible, but
not destroy the wonder, for when we have
explained the wonderful, unmasked the
hidden pattern, a new wonder arises at how
complexity was woven out of simplicity.

Herbert Simon
The Science of the Artificial

Bad systems are complex, hard to change, hardware and software dependent, and monolithic. Moreover, they are large, costly, and time-consuming to develop.

Unfortunately, knowing what a bad system looks like does not necessarily provide a clear guide for developing a good one. Over the last twenty years, many methodologies were developed that aimed at avoiding the creation of bad systems, but not until the early 1970s did approaches appear that defined what a good system should look like.

I still recall that when I first read about top-down structured systems design, it sounded "right." In particular, I was struck with the idea of building systems using only a few basic structures. Since all the basic sciences have developed around the systematic use of a few concepts or structures, the use of these structures to produce a system seemed to be an extremely promising approach.

This restriction to a few basic structures led to top-down (hierarchical) design, and that led to data-structured design. As designs based on hierarchical structures were developed, the connection between the structure of data and the structure of the programs or systems that processed the data became apparent.

It also became clear that systems hierarchies could be modeled after the aggregate structure of the outputs. Data-structured design yielded what might be called cycle-structured systems designs, one of the major ideas that served as the basis for much of **Structured Systems Development.**

Equally important in **Structured Systems Development** was the concept of output-oriented design. The basic idea behind output-oriented design, like the idea behind data-structured design, is simple enough: start with the output and work backward. Since the time of the Greek philosophers, organizations have been explaining functions in terms of goals, and systems in terms of outputs. Output-oriented design was simply a formulation of that idea applied to systems design in a rigorous manner.

In retrospect, the significance of output-oriented design was obvious; however, it took a lot of trial and error (the application of heuristics) to arrive at a useful definition. In fact, to attain a clear understanding of output-oriented design, we had to go through an experimental phase with top-down design.

During the period of enthusiasm over top-down concepts, a number of unforeseen and undocumented difficulties were encountered. The top of the hierarchy was easy enough to find (one simply wrote the "top"), but the next level of the hierarchy was not so easy to specify. Many top-down solutions, including a great many that worked, seemed arbitrary. It became clear that if we were to teach others how to design systems in a systematic way, something better than top-down design was needed.

OUTPUT-ORIENTED DESIGN

It was at this point that the idea of using the outputs to aid in defining the system began to take shape. Determining the outputs and working backward to design the data base and inputs was intuitively appealing.

Moreover, it was in keeping with many ideas inherited from traditional systems design. Rather than rejecting previous systems approaches as top-down design had, we came to appreciate and use approaches that had been perfected in the early stages of data processing but had since fallen into disfavor.

Output-oriented systems design is a very old concept indeed, but for various reasons, many people thought it was inferior to more "advanced" or "modern" approaches that taught it was better to start at some other point. For example, many approaches recommend concentrating on the processes of the organization, others on the data collected (input), and others still on the data base. This happens because output-oriented design has its difficulties, and many analysts and users prefer to solve their problems in a different fashion.

Solid Method for Designing Systems

Over the years, my colleagues and I probably have encountered every possible criticism to output-oriented design that could be raised. Moreover, the criticisms came from our friends as well as from our enemies. As we analyzed the criticisms, though, we discovered that they fell into two categories--those that were outside the scope of systems design proper, and those that were irrelevant. With all its difficulties, output-oriented design is the only solid method for designing systems, especially from the standpoint of systems theory. Only an output-oriented method of design can ensure that:

- all (and only) the data required is captured and stored;

- only necessary processes are used;
- the scope of the systems effort is clearly delineated.

To be fair, output-oriented systems design is open to certain criticisms. For example, it is dependent upon users' perceptions of their needs. It is also considered to be less flexible in providing for ad hoc requests for data that has not been used previously within the system.

But its strengths more than make up for its failings. Output-oriented systems designs are: complete, minimal, efficient, understandable, and generally suited to their definition.

The benefits of output-oriented design are hard to question. Even in those cases where individuals have taken the recommendation to "define the outputs first" as gospel, the results have generally been good. **Structured Systems Development** preached the point: *Define the outputs; make sure that both you and your client know where you are going from the outset. Then, after you have defined the outputs, make sure you develop the system so it will produce just the outputs required.*

This approach has streamlined the systems development procedure enormously. Beginning systems analysts were given a focal point; since the analysts knew what they were looking for (the outputs), they were able to deliver systems that did what the user requested.

But there were still problems connected with output-oriented design. What about the client who doesn't know what he wants? What about future outputs that may not be defined at the time the system is originally specified? What if the user makes changes? The list of objections to starting with the outputs is not inconsiderable.

The main problem most systems professionals and users run up against, however, is knowing when all the mandatory outputs have been collected. If you are certain that you have a complete set of outputs, or are able to obtain them, then the process of systems design becomes predictable. Today we recognize that in the systems development life cycle, the design phase (systems design) is preceded by planning and requirements definition phases and followed by construction, testing, and installation phases.

If you make this distinction between phases, then it becomes clear that the conflict over the the importance of outputs to design is really a confusion over the various phases. It is imperative, if you are to produce a good design, that the outputs be defined. That does not necessarily mean that you start the development process with a definition of the outputs, however; it does means that by the time you have finished the requirements definition phase, you must have a good definition of the principal outputs that the system must produce.

Consider the following analogy: if the process of designing a system is thought of as a mathematical equation, then the outputs become the independent variables, and the data base, the inputs, and the processes become the dependent variables. From the outputs and, of course, from the calculation rules for computing the outputs, everything else can be determined.

Output orientation leads to a rigorous design methodology. Since data does not materialize magically, everything that appears on an output must be put into the system as an input, or generated from some input.

By working backward from the outputs required, it is possible to determine exactly the minimal data base and inputs for the system.

Agreement with the Real World

It is desirable to have a minimal system. It's more important, however, that the system produces answers that agree with the real world. Output-oriented design gives this capability to us. If accurate and consistent data is to be obtained, then the data must be systematically and frequently put out and used. Otherwise, the answers we get will be wrong.

● **Data that is not used will not be correct.**

This rather extreme statement results from a significant application of systems theory (cybernetics) to information systems design. Indeed, systems theory can be used to show that data quality is directly related to data use, not to data collection, processing, or storage. Even if every precaution is taken in the collection, editing, and storage of data, if it is not used (or is infrequently used), the data will become increasingly inaccurate because, over time, there will be no feedback mechanism to ensure that it will be changed to reflect the real world. Therefore, if data bases are to describe reality accurately, the data must be constantly reviewed and updated.

More than any other approach, therefore, output-oriented design has the advantage of producing minimal systems in which the data will be maintained correctly. In addition, it allows a precise definition of the scope of a system. If the output is used as the key to design, then the designer has a good way of knowing when a system activity has been completed. Other systems design approaches (process, data base, input) are open-ended; they afford no way for the analyst/designer to be sure that all the inputs or processes required have been developed.

The more you look at it, the better output-oriented design appears. But it is not simply a case of being right where lots of others have been wrong. Those who chose instead to look at the processes or data base or inputs had reasons to do so, perhaps as a result of a personal experience, or of some difficulty in applying output-oriented design to specific problems.

Information systems support business (organizational) systems. The outputs of a good information system serve the organization in producing better goods and services, and in making better decisions. As more experience was gained in structured design, a considerable difference was recognized between a system developed to provide an arbitrary set of outputs specified by an uninterested user, and a system developed on a complete set of mandatory outputs defined by a rigorous requirements definition process.

The risk factor in building systems based on the output-oriented model is directly related to the certainty that all mandatory outputs have indeed been defined. As one goes from existing systems where outputs have already been determined to newer systems where they have not, the uncertainty, and therefore the risk, associated with developing the system using the output-oriented model go up.

STRUCTURED REQUIREMENTS DEFINITION

It became increasingly important to develop a sound approach that would direct the user to the discovery of the correct set of outputs for new or unfamiliar systems. The catalyst in this instance was a project our organization had undertaken to develop a tax collection system for a governmental agency. In the attempt to obtain the definition of the required outputs, a number of major ideas jelled, and several new tools were either developed or perfected.

Several questions provided a jumping off place for developing a systematic method of defining a system. If you don't start the development process with the definition of outputs, where do you start? What do you do on a new project or with a new client when you do not have a good idea of what the system is intended to produce? How do you get a reasonable start?

Through trial and error, and the application of many of our basic theories to the development of the systems process, we began to develop an approach that started with a definition of the scope of the project, and continued through the definition of the functional flow, the decision/control system, the functional subsystems, the tasks, and the outputs to be produced.

Over time, the process that evolved has been considerably modified. We came to understand better what had been done on the project from both the practical and theoretical points of view. The result was what we now call **structured requirements definition** .

In the same way that structured programming was the catalyst for the development of structured systems design, structured systems design was the catalyst for the development of structured requirements definition.

The Emergence of Structured Planning

But, even with a good approach to requirements definition, it is still possible to define a system that is not correct, i.e., one that does not address the real problems of the organization. Identifying the "right" system or application to be addressed is the goal of **structured planning and analysis**.

The more we understand about the nature of systems science, the more we know about the proper functions of each of the phases, and the roles of the professionals involved. We are increasingly aware of the necessary interfaces between the information systems planner (systems analyst) and the basic business or organizational planner (management analyst). This is not altogether surprising since management analysts and systems analysts are both engaged in the same process: to understand, describe, and solve organizational problems. Many of the most severe management problems have to do with the lack of accurate, timely information.

In addition, management analysts and systems analysts both have to deal with complex situations and with human factors. Management analysts bring to the table a long history of eliciting problems, while systems analysts bring an understanding of complex systems. Indeed, each can learn from the other.

Analysis and Synthesis

In the early days of what has been called the structured revolution, some of the naive presentations of top-down design promoted a code-a-little, test-a-little mentality. Unfortunately, this particular view of

top-down design ignored a major distinction in logical
thinking, i.e., the distinction between understanding
the pieces of a problem, and putting the pieces back
together in a solution.

As in so many areas of human thought, the Greeks
understood the importance of this distinction. They
had two words, one (analysis) for breaking things into
pieces, and another (synthesis) for putting the pieces
back together. In practice, this distinction is often
overlooked, and there is a tendency to go directly from
the problems to the solutions. We fail to take time to
understand exactly what our problems are in the rush to
implement solutions. The results are predictable.

The first papers on top-down design (really top-down
development) encouraged more thoughtful design by in-
sisting on a careful hierarchical design, yet discour-
aged complete understanding by attempting to implement
(synthesize) in parallel with analysis. And in the
early days of Structured Systems Design, we were guilty
of confusing the planning, requirements definition, and
design activities.

Teaching, consulting, and designing systems projects, though, have the effect of testing concepts and
theories. Feedback is certainly important in all human
learning experiences and there is perhaps no more
intense and difficult form of feedback than teaching,
especially when it includes follow-up, to see how the
principles, once taught, are applied. Somewhat fortui-
tously, we have been able to test most of these ideas
time and again against real problems in real organizations. Where we have been willing to listen, the
feedback from these experiences has kept us from fall-
ing into many of the more difficult pitfalls.

A variety of hierarchies are found in nature. Science is forever discovering that the hierarchies of the natural world are not accidental, nor do they occur at arbitrary points. A hierarchical subdivision in nature always matches some important aspect in the physical universe.

The same is true, we have discovered, for programs and systems. The hierarchy of a well-structured program naturally falls along the same lines as the structure of the data it is processing.

If there is a natural structure for programs and systems, why not for methodologies as well? Points of natural division occur in the life cycle of a system, and whether they are called by one name or another is ultimately of little consequence. What matters is that we do the right things in each phase.

New Tools Mean New Solutions

In the last few years, most of the controversies that have raged among the various systems thinkers have been the result of misunderstanding the viewpoint of a particular thinker. These misunderstandings were allowed to obscure our thoughts because we lacked adequate tools to describe and model systems. Thus many of the most important recent systems discoveries have been directly connected to having effective ways to express what were previously obscure relationships.

New tools make it possible to come up with new solutions, and these new solutions provide the insights to help us develop other new tools. But tools without a basis in theory are often a means of technical virtuosity where the objective becomes subordinated to the niceties of execution or to the skills of the executor.

Many of the long-term problems with programming and programming languages stem from a lack of theoretical underpinnings. Therefore, in any sound systems science there is the need for consistent improvement of all elements: theory, methodology, and tools.

Methodology and Theory

Human beings have a love/hate relationship with rules. On the one hand, they express their love of freedom and of being unhampered by restrictive rules; on the other, they must have rules to move from one level of civilization to another. Rules and methods are like habits: they allow us to solve a problem and not have to think about it again.

But methodology without theory becomes rigid. It lacks the ability to grow, change, adapt. The more complete and successful the method the more likely this is to happen. Without an appreciation for theory, however, human beings are apt to focus on superficial differences rather than on essential similarities.

A good theory is apt to generate a number of methodologies, each one appropriate to a given set of circumstances. Over the last few years, a great many theories, methodologies, and tools have been propounded. This is altogether natural, and from the standpoint of the propagation and dissemination of ideas, even desirable. If we are to gain a better understanding of systems development, we must provide and encourage mechanisms whereby vital concepts are widely discussed and tested.

Thanks to the work of men like Warnier, Dijkstra, Constantine, Ross, Jackson, and literally thousands of others over the last few years, the science of systems building is rapidly taking shape.   As it does, the important thing to remember is not **who** is right, but **what** is right.   The remainder of this book is an attempt to define what is right given the knowledge we have today.

Fundamental Concepts/3

****

Study without thought is vain,
Thought without study is dangerous.

Chinese proverb

When practitioners of vital professions are trained, that training must be carried out in a careful and specific manner. Nowhere is this more true than in life-and-death areas such as medicine or commercial air travel. By observing how practitioners are trained in these critical fields, systems science professionals can learn how to develop better training methods in their own field.

Consider teaching surgeons their craft. We begin by teaching them mathematics, physics, chemistry, biology, and anatomy. Next, we let them experiment on objects that don't count (or can't sue), i.e., frogs and cadavers. Then, we let them practice under the guidance of competent professionals. And, finally, we let them solo, but aided by trained assistants (nurses, etc.) and armed with the best possible tools.

Almost none of this rigidly structured professional training applies to training systems analysts, designers, or programmers. In general, we consider a programmed instruction course in COBOL or FORTRAN as basic training. We let beginners learn on live projects, rather than on test projects, and provide little, if any, technical guidance. To add insult to injury, trainee systems analysts are often given trainee programmers to help them. Finally, the tools we provide are minimal, or even harmful, considering our programming languages and operating systems.

If we are to develop skilled systems scientists, we must begin to emulate the true professionals. In a manner of speaking, this book is guided by this idea.

We want to approach the training of systems scientists in a structured manner.

This chapter, then, represents the mathematics, physics, chemistry, biology, and anatomy of information systems. There is nothing difficult contained herein, but the subjects are exceedingly important if we are to understand our craft and practice it effectively.

SOME BASIC MATHEMATICS

In school, my major fields were mathematics and logic. I can't say that I was a great mathematician or logician, but I did develop a certain understanding of the subjects, as well as a large amount of respect for their power and applicability.

When I entered the programming field, I began to think there should be a direct way to apply mathematics and logic to the development of correct programs and systems. It seemed the natural thing to do. Unfortunately, I was never able to make my programs logical. Perhaps I was confused by the phrase "program logic," as though programs were supposed to be logical. Moreover, none of the mathematicians cum programmers I knew were consistently successful in turning their knowledge of logic into creating correct programs either.

I had the same conceptual problem in applying "systems theory" to the building of systems. I knew that an entire branch of science, called cybernetics, dealt exclusively with the study of "systems." Although cybernetics was vitally interesting to individuals involved in building control machines, its concepts were difficult to apply to the information systems I was building. I knew that such concepts as "feedback" and "control" were important in all systems, but I could not find an easy method of applying them.

For me, it was the ideas involved in structured programming that first tied mathematics together with program and systems design. I knew intuitively that the basic structures of structured programming (sequence, alternation, repetition) were directly comparable to the basic structures of mathematical logic (sequential proof, the decision rule, and quantification). And all finite mathematics could be developed using only these basic building blocks.

With the application of basic structures, information systems science was at last getting somewhere. Programs and systems could be viewed as "mathematical objects," to use Harlan Mills's phrase, and were subject to sets of rules and evaluation techniques developed over thousands of years. Programming could now be viewed as an operational definition of mathematical logic. This new foundation made it possible to move from the realm of personal preference to the realm of proof as a means of evaluating programs and systems.

Data structuring, i.e., the development of program logic from the data structures, went even further in eliminating personal idiosyncracies from design. Data and process had always been considered necessary to the development of good solutions; the discovery that a natural symmetry existed between the structure of the process and the structure of the data it processed was a major theoretical breakthrough.

Warnier's work added further conviction that systems science was moving in the direction of developing more rigorous foundations. Warnier's basic point of view was that all programming problems could be resolved in terms of set theory. His method of organizing data and actions into hierarchically organized sets and subsets allowed the direct translation of problem statements into problem solutions; moreover, this process always produced structured solutions.

With time and experience, the connection between information systems and set theory has become even more apparent. So, too, has the dependence of good systems design and operation upon the fundamental cybernetic ideas of feedback and control.

Static View--Sets, Subsets, Elements, and Attributes

Many difficulties in designing information systems revolve around an inability to state a problem completely and precisely. The logical tools provided by set theory make precision much easier.

A **set** is simply a collection of similar objects or things. Each object is called an **element**.

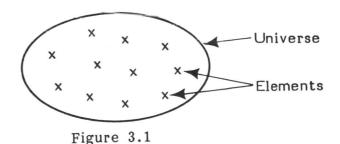

Figure 3.1

A bank, for example, has customers or clients. If we describe a set of customers who do business with a specific bank (Fig. 3.2), we have identified a set of elements (customers) that belong to a certain set or **universe** (bank).

Elements have **attributes** or **properties** that serve to

describe them.   A customer may have a name, an address,
and any number of other attributes.

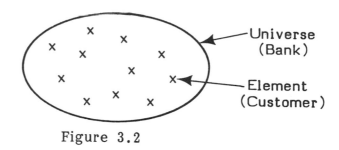

Figure 3.2

Within a set, it is possible to group elements into
subsets according to their attributes.   For example,
we may group the customers of a bank into subsets based
on whether the customers have a savings account attribute or a checking account attribute, or both.

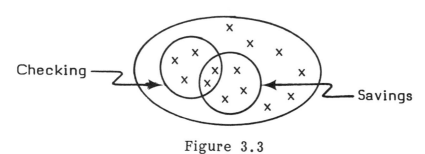

Figure 3.3

The Relations Between Sets

It is possible to define complex operations in a
simple manner using only a few basic relations.   For
instance, **intersection, union,** and **complement** are all
relations defined between sets and subsets.

The **intersection** of two sets is defined as a subset
of those elements common to both subsets.   The **union**

of two subsets is defined as a subset containing those elements that belong to one subset or the other, or to both. The **complement** of a subset is the subset of all the elements within the universe that do not have the attribute that characterizes those elements in the original subset. Figure 3.4 shows examples of these relationships.

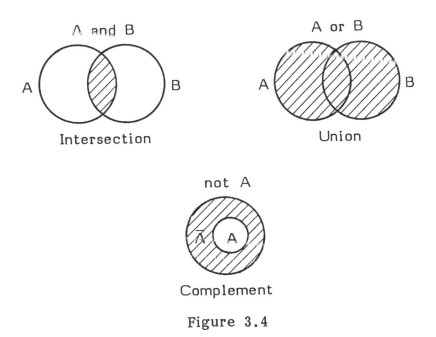

Figure 3.4

Set intersection and union are possible only for subsets defined by attributes that are independent of one another. If one subset is defined so that it falls completely within another, as in the case of a special checking subset within the checking subset, then the first is called a **proper subset** of the second. Proper subsets are particularly useful because they allow us to build hierarchies from sets and subsets. A **partition** of a set divides its elements into two or more proper subsets that are also exhaustive (no element left out).

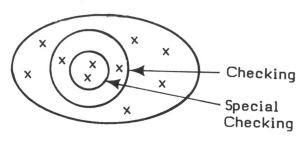

Proper Subset

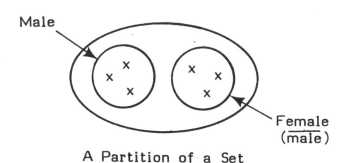

A Partition of a Set

Figure 3.5

Set representation gives the systems scientist a powerful means of modeling the real world. If the real world can be depicted in terms of sets and subsets, then it is possible to apply the rules and theorems of set theory to the solution of information problems.

Consider the following problem from **An Introduction to Decision Logic Tables**[1] by Herman McDaniel:

[1] McDaniel, Herman. **An Introduction to Decision Logic Tables** (Rev.ed.). Princeton, NJ: Petrocelli Books, 1978.

A steamship line has a reservations and ticket counter where customers make reservations and purchase cruise tickets. If a customer requests cabin-class accommodations and a cabin is available, issue a cabin-class ticket and subtract one from the total number of cabins (so as not to oversell the capacity of the ship). If cabin-class accommodations are not available, place the customer's name on a waiting list for cabin class. If the customer requests tourist accommodations and they are available, issue a tourist-class ticket. You must also remember to subtract one from tourist availability. If tourist class is requested but is not available, place the customer's name on a waiting list for tourist class. For the purposes of this problem, if the class initially requested is not available, there is no possibility of assigning an alternate class.

This particular problem can be described in a logical, graphical manner in terms of sets, elements, subsets, and attributes. In this particular case, the universe is the steamship line and the elements are customers requesting tickets. The attributes are:

- customers requesting cabin class ($\overline{\text{CR}}$)
- customers requesting tourist class (CR)
- customers for whom cabin class is available (C)
- customers for whom tourist class is available (T)

The set diagram (also called a Venn diagram) of the problem appears in Figure 3.6.

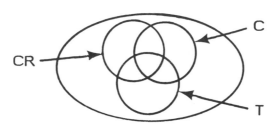

Figure 3.6

This diagram is particularly important because it shows all the logical possibilities posed by the problem statement. In this case, we must deal with eight possible, unique subsets. The diagram allows us to identify each unique subset, and from the problem statement we can determine what must be done for each one.

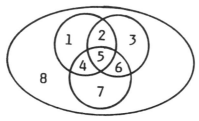

Figure 3.7

The various subsets, with the attributes and the actions for each subset, can be used to develop a truth or decision table (shown on page 33).

Physical Realizations

To take advantage of the power provided by set theory, one must be able to develop physical representations of the various logical entities involved. The

Subset	Attributes	Actions
1	CR.$\overline{C}$.$\overline{T}$.	Place on waiting list for cabin class
2	CR.C.$\overline{T}$.	Assign cabin class, subtract 1 from cabin class list
3	$\overline{CR}$.C.$\overline{T}$.	Place on waiting list for tourist class
4	CR.$\overline{C}$.T.	Place on waiting list for cabin class
5	CR.C.T.	Assign cabin class, subtract 1 from cabin class list
6	$\overline{CR}$.C.T.	Assign tourist class, subtract 1 from tourist class list
7	$\overline{CR}$.$\overline{C}$.T.	Assign tourist class, subtract 1 from tourist class list
8	$\overline{CR}$.$\overline{C}$.$\overline{T}$.	Place on waiting list for tourist class

computer, for example, is actually the outgrowth of a physical device invented to perform the basic mathematical functions. We call a physical solution to a logical problem a realization.

A **realization** is a representation or model of a mathematical or logical set in physical or operational terms. A data file can be thought of as an information set. A customer file within a bank might have one record per element (customer) and, within each record, separate data attributes (name, address, checking, savings) that appear as data fields.

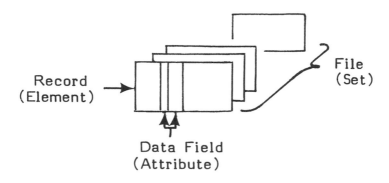

Figure 3.8. Physical Realization

Every logical situation has, potentially, an infinite number of different physical realizations that are equivalent, in the sense that each produces the same result. From a theoretical standpoint, it does not matter whether one represents a logical element as a punched card, as a set of bits on a magnetic tape or disk, or as holes in an optical disk. Nor does it matter how the data is encoded, as long as the relations of the logical set diagram can be represented unambiguously in the physical solution.

In the early days of data processing, a great deal of time and attention was given to the "best" or "most efficient" means of physical representation. This was natural because of the high cost of storage and processing. But with the advances in computer hardware, more attention has been devoted to the logical solution. Both areas are important; however, the logical solution is primary and the physical solution must always wait

---

upon the development of a correct logical one.

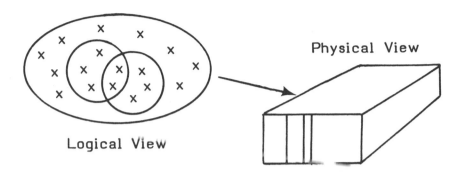

Figure 3.9

For the sake of simplicity, we will use the following correspondences:

Logical Set	=	Physical File
Logical Element	=	Physical Record
Logical Attribute	=	Physical Data Element

This set of correspondences will allow us to make a simple transformation from the logical to the physical whenever it is necessary.

MAPPINGS

If mathematics defines the ideal static world, then physics and chemistry can be thought of as defining the dynamic world from simple idealized components. In the same way that mathematics provides useful tools for physics and chemistry, set theory provides tools for the systematic development of complex processes and systems.

Logical representations of sets, subsets, elements, and their attributes allow us to describe a certain class of problems; however, systems scientists are, for the most part, interested in dynamic relations between sets. Such a dynamic relation between sets is called a mapping.

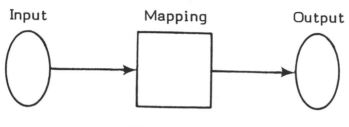

Input   Mapping   Output

Figure 3.10

A **mapping** is a transformation of one or more sets into others. In systems terms, the original set(s) is called the **input(s)**, the final set(s) is called the **output(s)**, and the mappings themselves are called **processes**, **programs**, or **transforms**.

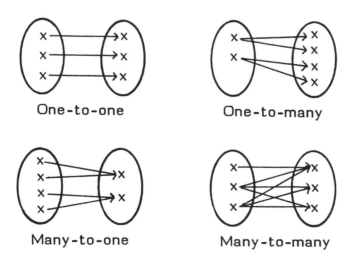

One-to-one   One-to-many

Many-to-one   Many-to-many

Figure 3.11.   Kinds of Mappings

Types of Mappings

There are three kinds of simple mappings: **one-to-one, one-to-many**, and **many-to-one**. There is only one kind of complex mapping: **many-to-many**.

At an abstract level, many systems problems can be formulated in terms of mappings between information sets. For instance, the customer set of the bank can be mapped into a subset of those customers who are residents of California.

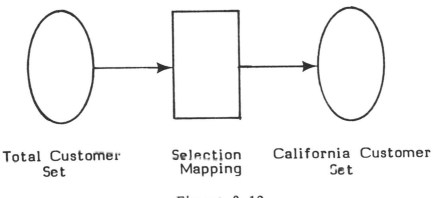

Total Customer   Selection   California Customer
Set   Mapping   Set

Figure 3.12

In Figure 3.13, we show the mapping of two input sets, the current customer set and the customer change set, into a new customer set through some mapping (update) process.

A program to edit data is a mapping; so is a sort routine. Since they are both mappings, they can be defined logically in terms of operations (transforms) on sets and subsets of data. (Mappings may be as simple as a program to copy one set of information to another, or as complex as a military command and control system.)

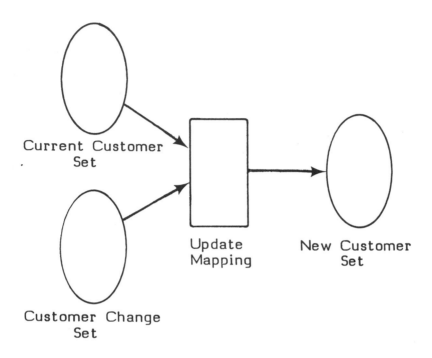

Figure 3.13

From a theoretical standpoint, the real power of mappings results from their use in series. A complex problem can be attacked by making a series of logical passes over the data, doing something very simple in each pass.

Most major problems arise from the use of many-to-many mappings--"structure clashes," as Jackson calls them. For example, we have difficulty getting correct headings on report pages, for example, because the logical structure of the data does not map neatly onto the physical structure of pages. The same is true of the systems problem encountered in attempting to relate weeks and months, since the relationship between weeks and months is many-to-many.

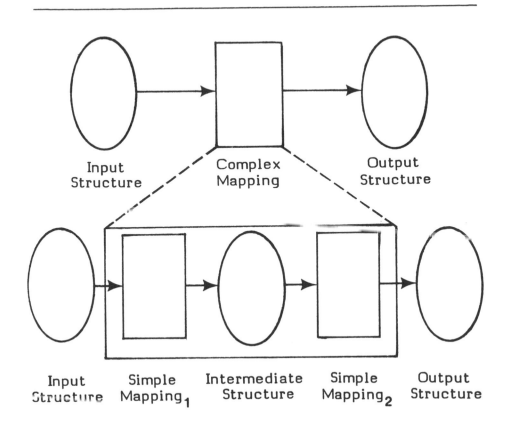

Figure 3.14

Having a systematic solution for many-to-many problems is critical. Basically, the strategy is to try to eliminate many-to-many mappings entirely if possible. Since simple mappings yield simple and workable solutions, the approach for dealing with complex mappings is to break them into a series of simple ones. This entails finding some common element that occurs in both the input and the output sets.

Remember: There is no problem so big that it
can't be run away from.
                                            Peanuts

Physical Realizations of Mappings

It does not matter how one represents information sets and mappings because they are, in fact, all equivalent. Whether a set is represented as an oval, or as an arrow, or as a tape or disk, it is still an information set. Whether a mapping is represented as a box or as an oval or simply as a line does not matter; it is still a mapping. What matters is having a means of representing information sets and the mappings between

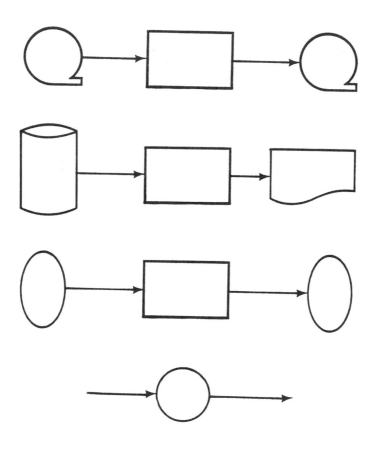

Figure 3.15.   Graphical Representations of Mappings

them. Getting your viewpoint clear is nearly half the battle.

Just as there are an infinite number of different physical realizations of a set, there are also an infinite number of physical realizations of a mapping. A mapping can be a FORTRAN, COBOL, or PL/1 program, a canned procedure, or part of the hardware or firmware.

The practical applications of mappings are enormous. The nature of successful large-scale design is predicated on breaking the overall problem into a series of phases, or passes. Consider the following example:

> A computer manufacturer wants to develop a COBOL compiler for a new class of computers. This compiler will scan the source, which may be on an input file, and produce object programs in a form suitable for the systems loader to process.

Compiler development has been an on going activity for a long time. The basic mappings that have evolved over the last 25 years are shown in Figure 3.16.

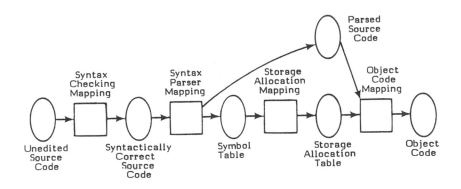

Figure 3.16

In this particular case, the mappings from one step to another are complex in themselves, and a syntax checker may run many thousands of lines of computer instructions. The description shown in Figure 3.16 is fundamentally correct, however, and for the most part, all current compilers operate in this way.

The importance and popularity of the various forms of data flow diagrams, and data-flow based structured design techniques (Constantine and Yourdon, Ganes and Sarson, Weinberg, DeMarco, etc.), is witness to the power of simple mappings in solving complex problems. At each level of design, mappings enable the systems scientist to develop a deeper insight into the logical characteristics of the problem.

THE CYBERNETIC VIEW: FEEDBACK AND CONTROL

Because mappings are powerful, there is a tendency to stop at that and consider everything as a mapping. Much technical literature treats systems simply as large programs--as mappings from inputs to outputs. This particular view is often captured in diagrams called input-process-output (IPO).

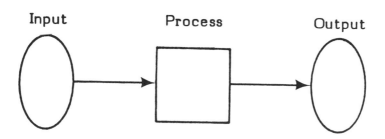

Figure 3.17.   IPO Diagrams

The approach has a great deal of appeal, in part because it is so simple (too simple, in fact). But we will show that a system is more than a one-time mapping of a set of inputs into a set of outputs; true systems are distinguished by their attributes. From a logical standpoint, true systems are:

- self-controlling,
- self-correcting,
- goal directed, and
- persistent.

By introducing these factors into the discussion, we have moved beyond the simple IPO model. But if the input-process-output model does not adequately describe a system, then what does? As a starting point, we need to look at the models developed by cybernetics.

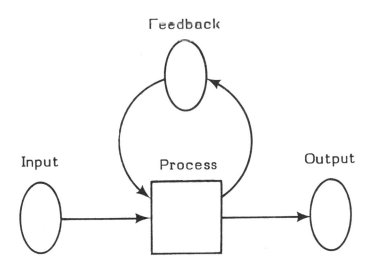

Figure 3.18. A Feedback/Control System

A toaster, for example, contains a control system. A color setting on the toaster controls a process, regulated by a timer, that determines the "darkness" of the toast. A fuse also acts as a control mechanism. An electrical device will operate until the capacity of the fuse is exceeded.

The control systems described above are called **open systems**. They can control behavior, so they are self-controlling, but they lack the capability for self-correction. Self-correcting systems are called **closed systems** (see Fig. 3.18).

In a sense, a **feedback/control system** produces output from input, but it does considerably more. It attempts to produce output that is within certain limits.

Let's consider the simplest form of feedback/control system (using a control diagram)--a heat thermostat.

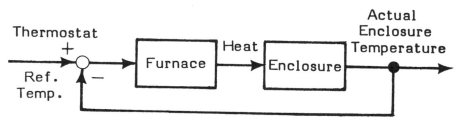

Figure 3.19

In this particular case, the line going back literally "feeds back" to modify the generation of the next cycle of input cycles. The output of the "heating system" is a level of heat within certain predefined limits. The effect of this can be seen by charting the temperature over a given period.

The output temperature is set for 72 degrees. The system turns on the heat if it is more than two degrees below the desired level, and turns it off again when the temperature rises more than two degrees above the desired level. The desired temperature is a goal, and in cybernetics, it is called the **reference signal**.

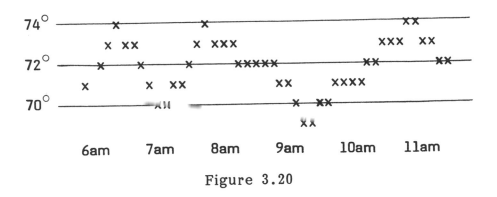

Figure 3.20

Goals (reference signals), cycles, feedback, and control are important characteristics of all good systems. Let's look at a simple accounts receivable system from a control systems standpoint.

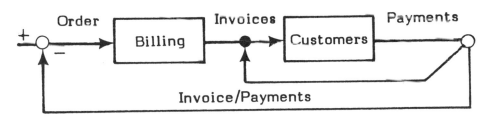

Figure 3.21

In this system, the ratio of invoices to payments is used to control the acceptance of orders. If the ratio of total invoices to unpaid invoices becomes too high, the feedback signal tells the system to restrict the input, and the acceptance of orders to reduce the ratio of unpaid orders. Thus the rate of payment of invoices controls the input of orders to the accounts receivable system in much the same way that a rise in temperature controls the input of heat to the heating system.

The major factor in feedback/control systems is their self-correcting behavior. Since they do control their own behavior, they tend to be highly reliable. The day-to-day business world, however, has had little appreciation of the importance of feedback and control systems. This is partly because cybernetics is such a new field, but also partly because it has been applied largely in technical areas.

> Feedback, one of the most fundamental processes existing in nature, is present in almost all dynamic systems, including those within man, among men, and between men and machines. However, feedback concepts have been utilized almost exclusively by engineers. As a result, the theory of feedback control systems has been developed as an engineering discipline for analyzing and designing practical control systems and other technological devices. Recognition that this theory is directly applicable to formulating and solving problems in many other fields is becoming widespread, but its use has been limited because of its heavy orientation toward technological applications.
>
> Joseph DiStephano, et al.
> **Feedback and Control Systems**

Organizations are, in large part, feedback/control systems. Alfred Sloan, the man who put together General Motors, has been quoted as saying that he could "run GM on just six ratios." That remark reflects a practical application of feedback/control. Cybernetic controls in business often take the form of ratios, e.g., return on investment and liquidity.

Although feedback and control can be applied in many situations, it is a recent phenomenon to think of an organization, or even of economies, in systems (cybernetic) terms. Systems scientists are also beginning to take such thinking seriously and apply it to building information systems.

The differences between systems developed using cybernetic concepts and those developed using traditional approaches can be considerable. Management information systems can be developed empirically by surveying managers and asking them what they want to know, or they can evolve from a scientific analysis of the business (functional) systems they support and the feedback/control required to ensure their successful operation. In the first case, the quality of the system developed is strictly limited by the skills and perceptions of a given set of managers and users at a specific time. In the second case, the quality of the system is based on the fundamental business functions of the organization.

The characteristics of good feedback/control systems are now well known. And we are seeing that the application of feedback/control principles to the design of management information systems can have a significant effect on the end product.

Beyond Simple Feedback and Control

Higher-level systems are not only self-correcting, they are also self-actualizing, i.e., they constantly adjust their goals to meet changes in the environment.

In nature, for instance, every living thing can be seen as a complex system; many, if not all, organisms attempt to modify their environment, to control their world. Animals and plants are constantly adjusting their objectives to meet changes in climate, food supply, etc. The more automatic the basic functions, the more the organism can busy itself about other, presumably more important, matters.

What is true for biological systems is equally true for management information systems. A perfectly rational investment policy may change dramatically when the interest rate jumps from 10 to 20 percent. An energy policy formulated when gasoline is 30 cents per gallon may have to be altered considerably when gasoline is $1.50 per gallon.

The heating system in Figure 3.22 adjusts its outputs in a two-stage manner. The system will not only control the temperature within a few degrees of a specified reference level (desired temperature), it will vary the reference level, depending upon the time of day.

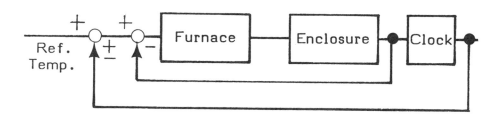

Figure 3.22

This results in a more complex behavior pattern (Fig. 3.23).

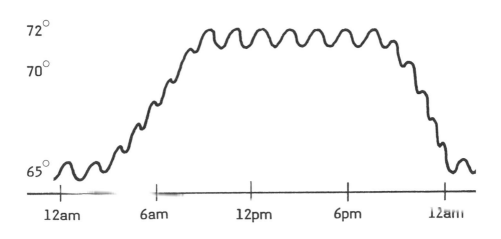

Figure 3.23

We have traded simplicity for economy. By raising and lowering the reference temperature to meet differing requirements, fuel economy is improved. This is exactly what biological systems do. When the energy requirements of an animal are reduced (during sleep or hibernation) the animal's basal temperature is also reduced.

We could easily make this system more complex by adding the need to cool as well as heat. This could be done naively by simply turning on the cooling system whenever the temperature rises a few degrees higher than the reference point. (I have actually stayed in hotel rooms with individual heating/cooling units that operated on this principle.)

Naive heating/cooling systems maintain reasonable indoor temperatures over a much broader range of external temperatures. A heating system can only ensure that the range is within prescribed limits when the outside temperature is less than or equal to the reference temperature for the specific time. Such systems provide comfort at a high cost, however. By adding another input to sense the outside temperature, the heating/cooling system can raise or lower the temperature until it is within the reference band.

We speak of something as being automatic if we do not need to think about it. Automatic systems are self-correcting. The more automatic a system is, the more things it controls. In addition to heating and cooling, our system could also maintain humidity levels. To do so completely would require both a humidifier and a dehumidifier, but it could be done, and in such a way that for the most part we would live more comfortably.

The development of more complex feedback/control features within our information systems allows us to control basic, vital functions and to have more time to attend to the other, less predictable, changes in our world. The price paid for a more complex feedback/control system, however, is the necessity to sense and retrieve additional pieces of information. In addition to knowing what the current temperature is in the heating/cooling example, we need to know the time, the outside temperature, and the outdoor and indoor humidity, as well. The more complex the control system, the more there is to keep track of. Thus, the system must be able to remember.

Memory

A higher-level feedback/control system must deal with many inputs vital to its existence, i.e., it must have a memory capacity. Memory is required because of some difference between the input and the output. For example, the outputs may occur:

- at a different time than the inputs
- in a different form from the inputs
- in a different structure from the inputs

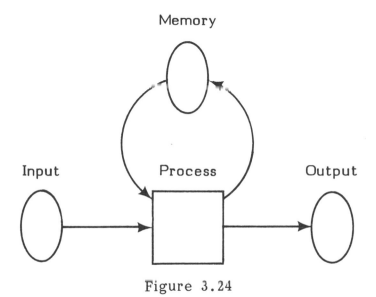

Figure 3.24

Our theoretical picture or model of an information system, then, is significantly affected by this need to remember. We have introduced a new dimension--time. This picture gives us a much more significant and useful model from a systems standpoint. All important information systems have some form of internal memory, so it is reasonable to think that a good logical model would have something analogous.

In Figure 3.24, the information set labeled "memory" must be able to remember feedback and control information, and it must provide storage to allow for resequencing, summarization, and time delays. All systems, even the simplest, have this requirement for memory.

We will lump memory, feedback, and control into a single set and call it the system's data base (Fig. 3.25).

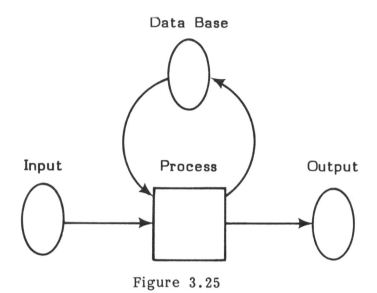

Figure 3.25

A **logical data base** (as distinguished from a physical data base) is simply all the information needed to produce the required outputs of the system. In a system, the principal outputs are always produced from the (logical) data base, and, in turn, the data base is produced (updated) from a previous (initial) version of the data base, plus any changes introduced by the inputs (see Fig. 3.26).

Cycles

In any complex system, many functions are being controlled at the same time. The importance, or difficulty, in controlling one aspect or another often varies considerably with time and circumstance. It would be a

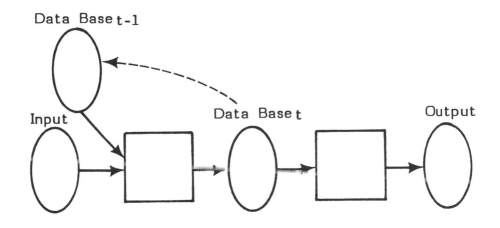

Figure 3.26

practical impossibility for systems to deal with the infinite range of variability in the world if some scheme for ordering the control were not used.

In the real world, both natural and artificial systems deal with complexity by seeking out repetitive patterns and then structuring themselves accordingly. Systems are repetitive by nature. Rarely is anything automated that is not, or cannot be fashioned as, cyclical.

Even allowing for cyclical behavior in the world, it is possible to build overly complex control systems. Good systems, both natural and man-made, avoid this difficulty by organizing their subsystems into hierarchies, where shorter control cycles are fitted neatly within longer ones in a precise manner.

If we view our heating system from the standpoint of cycles of control information we see, even in this simple example, two cycles of feedback--actual temperature and time.

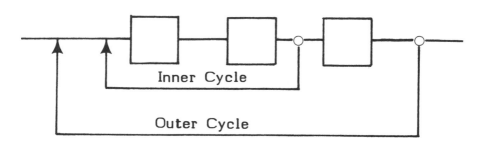

Figure 3.27

By recognizing this fact, a hierarchical feedback/control system can be built.

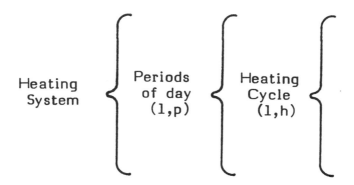

Figure 3.28

This system accomplishes the desired result, and does so very efficiently. Once a reference level is set, the system stays in the inner loop, regulating temperature until a change in the time period occurs. At that point it cycles back through the outer loop, resets the reference temperature, and enters the inner loop again.

The process can be carried even further. It would be possible, for example, to describe the inner thermostat of most hibernating mammals by adding yet another level of control--season--that would establish the reference temperatures for periods within each season.

Cycles and hierarchies of cycles allow systems to respond to a wide variety of changes in such a way that important aspects of the real world are controlled for the benefit of the system. Events occur in the real world in predictable cycles; therefore, the changes can be structured and dealt with one at time, rather than all together.

From information theory we know that if something is changing at some rate N (where N = 1/second, 1/month, 1/year, etc.), we need sample its behavior no more frequently than 2N. In fact, if we react too quickly, we tend to **overcontrol**, and if we sample any less frequently, we tend to **undercontrol**. The best systems, then, are those in which the system's behavior is precisely synchronized with its environment. For this reason, systems that have evolved over time are often better than those developed arbitrarily by systems analysts who did not have a clear understanding of the effect of cycles and information.

This does not mean, however, that the only way to develop good systems is by attempting to simulate the existing systems, either manual or automated. New knowledge of hierarchical feedback/control systems gives the systems scientist the ability to uncover the fundamental order of a system and to improve upon it.

Cycles in systems are related to the rate of change of important entities in the real world. If those changes occur at a predictable rate, then we can build systems to take advantage of that fact. Basic cycles are often related to the traditional calendar breakdowns, i.e., days, weeks, and months. Although these cycles are not universal (e.g., some systems are af-

fected by product or business cycles), they are significant enough to be reflected in many instances.

The structure of good systems, therefore, is neither
accidental nor arbitrary, but rather the result of
careful adaptation to the systems environment and
goals. In considering such cybernetic concepts as control, feedback, memory, cycles, and hierarchies, we are
not introducing new ideas into the building of information systems. These ideas have always been part of the
process of construction of every good system. We are
simply beginning to understand their importance and to
make their roles explicit.

STRUCTURED HIERARCHIES OF INFORMATION SETS

We have developed at least the beginnings, then, of
a theory of information systems based on sets, subsets,
elements, mappings, controls, feedback, memory, cycles,
and hierarchies. But we have yet to come to anything
truly new. Why, then, should we hope that structured
systems development will be better than previous approaches? The answer to this natural question is that
there is something that you have not been told: all
good systems are "structured," and we now know how to
demonstrate that structure.

Historically, systems scientists have been unable
to employ many of the concepts discussed in this
chapter because their basic training has been in dealing with simple sets of information. In the real
world, however, they must often work with systems made
up of information sets that are not simple sets but,
rather, are hierarchies of sets.

Hierarchies make dealing with the problems of complexity, efficiency, and, most importantly, control much easier. If one needs to do something complex in a reliable and efficient manner, then by all means the best way is to break down the tasks involved into a hierarchy of those tasks, where each task is responsible for specific outputs, and each is called on in a specific order.

To appreciate this process, look at the application of simple mappings to the development of a compiler, shown in Figure 3.29. In this case, a high-level mapping (converting source code into object code) is accomplished hierarchically through breaking down a single large mapping into a number of smaller ones. The output of one thus becomes the input to the next, with the various mappings connected in a controlled manner.

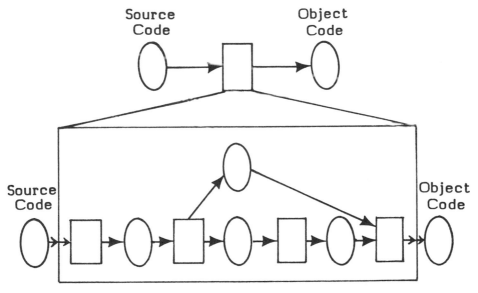

Figure 3.29

There is nothing very new or original in all this, either. Breaking down complex functions into simple ones is a well-known method of problem solving. In fact, all systems are developed this way. If a task is too complex for a single program to accomplish, then the process is divided among a series of programs, with the interconnecting inputs and outputs becoming data files.

We are now ready to define a structure.

A **structure** is a hierarchy of information sets in which the elements at each level are related (ordered) in terms of either sequence, alternation, repetition, concurrency, or recursion.

In other words, a "structured" system is one in which the pieces are related hierarchically to consistent data flows, and the control between the various pieces is one of the basic logical structures. This means it is possible to develop systems made up of layers of sets of data or actions, connected in simple ways.

One of the theoretical milestones of systems science was Bohm and Jacopini's proof that demonstrated it was possible to build a good program using only three logical means of construction: sequences, alternatives, and repetition of instruction. The consequences of Bohm and Jacopini's work was the assurance that a good solution could always be developed using only a structured, hierarchical organization. Since this kind of structure is simple, it is also easier to understand, manipulate, and control.

Equally important was Dijkstra's observation that the practical application of Bohm and Jacopini's theories was in the development of programs that did not need unconditional branches. This made it possible to produce "structured," or topologically nested, programs. Using only these basic connections, structured programs turn out to be naturally hierarchical.

Warnier and Jackson, working independently, came upon yet another major discovery. Each found that a natural relation exists between the structure of the data and the structure of the program that produced the data.

From all this, the germ of a new discipline developed--the theory of information structures. If there was such a thing as set theory, then why not a structure theory, where set theory would be the limiting case (a one-level structure)? It also occurred to the more radical thinkers that there was no good reason not to extend the structure upward as well as downward. Couldn't a system be thought of as simply one big hierarchy in which individual programs or procedures were merely modules? The answer was clearly yes.

Historically, the greatest precision in design had been at the program/procedure level. This had not been the case at higher levels, where procedures were related. Often at this highest level, the ordering was accomplished by human beings, and human beings do not require rigorous structuring. Therefore, it had been common practice not to provide any precise procedures for the highest levels of systems operation.

As it turned out, it was possible to structure the highest levels of a system just as precisely as it was possible to structure the lower levels. But even though the lower levels of the system could usually be structured using the sequence, alternation, and repetition structures first proposed by Bohm and Jacopini, it was not possible to treat elements at the systems level in the same way unless concurrency and recursion were introduced to the process.

The addition of concurrency and recursion has served to increase significantly the descriptive power of hierarchical structures without adding complexity. Used in the right fashion, then, hierarchical structures are extremely powerful tools for analysis, description, and development.

CONCLUSION

Systems science is involved with understanding the nature of information systems. The larger and more complex the type of system we develop, the more we need a theoretical, scientific basis. The methodology and tools used in structured planning and requirements analysis reflect many of the concepts we have already discussed.

It is difficult and time consuming to create workable solutions at all, but it is harder yet to create simple, correct, maintainable solutions to complex problems. Ultimately, however, only the simple, structured solutions will provide a firm basis for the construction of large information systems. Only simple, structured solutions can be managed and maintained over the many decades that our systems will have to operate.

Tools/4

****

There is something uncanny about the power of
a happily chosen idiographic language;
for it often allows one to express relations
which have no names in natural language
and therefore have never been noticed by anyone.
Symbolism then becomes an organ of discovery
rather than mere notation.

S.K. Langer
An Introduction to Symbolic Logic

Like a good surgeon, a good systems scientist needs good tools. Moreover, the tools should be appropriate to the tasks involved. In the case of systems definition they should allow us to show the following:

- Sets, subsets, elements, and attributes,
- Mappings,
- Feedback, control, and cycles,
- Memory, and
- Hierarchies.

In a sense, the structured revolution in systems design is a revolution in tools. Data flow diagrams, SADT®diagrams, Warnier/Orr diagrams, entity diagrams-- each has played a major role in recent advances. Our success with the use of these tools has led to the re-discovery of many older ones, such as Venn diagrams and decision tables, that are extremely useful if used correctly, but have been put aside in recent years.

The study of the history of technology reveals that tools usually evolve from a need to solve a particular set of problems, or to accomplish certain tasks. This is particularly true of structured tools. Over the years, out of the need to solve specific problems, we have crafted tools to aid us. Since not every tool is good for every purpose, it is important for the systems scientist to know what tools are available, and how to use them. He should know which tools are equivalent, and which are best in a given circumstance. Since tools are highly related to procedures, he should also understand how his tools fit in the overall picture.

In Chapter 3, we discussed information sets and structures, and tried to show their roles in the progression from static to dynamic to cybernetic views of the world. In this chapter, we will proceed in much the same way to consider tools that also represent static, dynamic, and cybernetic views.

Sets and Structures

Much of the pioneering work on the logic of sets was done in the nineteenth century. During this period, one of the major tools developed to describe sets was the **Venn diagram**. We rediscovered them when we discussed sets, subsets, elements, and relationships.

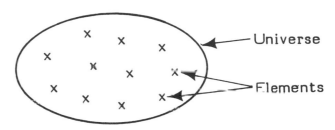

Figure 4.1. Venn Diagram

The Venn diagram is simplicity itself. Circles or ellipses are used to define sets and subsets (universes), and X's are used to define elements. (When working at an abstract level where it is only necessary to represent relations between sets and subsets, the elements are often omitted.) All the significant relations between sets (intersection, union, and complement) are shown graphically by the bounded areas within the diagram.

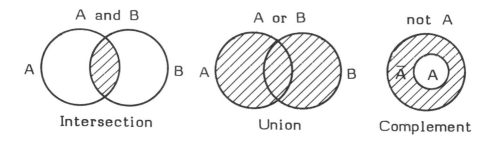

Figure 4.2

Venn diagrams can be used to explain logical relationships between various subsets of data even to
unsophisticated users. Suppose, for example, you were
assigned to develop a project management system for
data processing management. Let's further suppose that
you were having difficulty explaining the various
logical possibilities concerning critical projects. By
developing a Venn diagram, it would be possible to get
the user to see that he is not just interested in
projects that are in trouble, but more importantly, in
troubled projects that he can do something about.

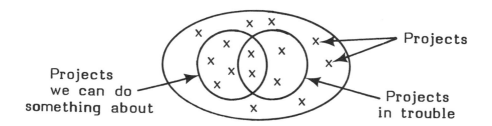

Figure 4.3

This diagram visually displays the projects that
cannot be helped. If management can grasp this important idea, then a great deal of wasted effort can be

---

avoided, and an information system developed that will indeed aid management in the solution of real problems.

The Venn diagram, as well as being valuable in communicating logical ideas, also has two characteristics common to all good structured tools: it is graphical, and it is simple. It is this very simplicity, however, that is the diagram's principal disadvantage. Venn diagrams only work for uncomplicated relationships. If the problem situation involves more than three independent subsets, the diagram becomes hopelessly difficult to draw. Moreover, such diagrams are useless in depicting hierarchies of sets (structures).

Warnier/Orr Diagrams

For complex analysis, we will use a far more powerful tool--the Warnier/Orr diagram. One can think of the Warnier/Orr diagram growing naturally out of the common formal definition of sets used in mathematics. In mathematics, a set is defined through the use of braces (curly brackets).

$$X = \{a,b,c,d,e\}$$

In this case, the braces represent the boundary (universe) of the set **X** that is made up of elements **a**, **b**, **c**, **d**, and **e**.

A Warnier/Orr diagram can be thought of as a set definition where the equals sign (=) and the right brace are dispensed with, and the elements are listed vertically instead of horizontally. But the other meanings are the same--**X** is a set, and **a**, **b**, **c**, **d**, and **e** are elements.

Figure 4.4

There is one major difference, however. Unlike traditional set theory that considers elements within sets as unordered, the normal case in structure theory is to deal with sets, in which the elements are ordered. In Figure 4.4, then, we would assume that elements **a**, **b**, **c**, **d**, and **e**, are ordered sequentially. (It is possible to use a Warnier/Orr diagram to describe unordered sets by using the concurrency operator "+," but this is the exception rather than the rule.)

Since information within a Warnier/Orr diagram is always described in hierarchical form, users find that the diagrams do not grow complex at the same rate as Venn diagrams when variables are added. More importantly, Warnier/Orr diagrams are capable of defining structures, i.e., hierarchies of sets, naturally.

Let's take as an example the definition of the data of an invoice or bill. One can think of an invoice as a set of information that is made up of:

- a subset of information that occurs one time at the beginning of the invoice (invoice number, invoice date, customer name, customer address, shipping address),

- a subset of information that occurs one
  time per product per invoice (product number, product description, number of units,
  price per unit, extension),

- a subset of information that occurs one
  time at the end of the invoice (sales tax,
  freight, invoice total).

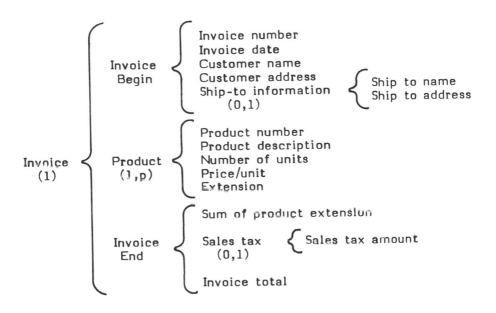

Figure 4.5

This is truly a hierarchy of information sets, in
which the immediate elements of the INVOICE SET are (1)
an INVOICE BEGINNING SET, (2) a PRODUCT SET for
each product on the invoice, and (3) an INVOICE ENDING
SET. In this case, the elements of the INVOICE SET
are, in fact, sets themselves.

By using Warnier/Orr diagrams, there is really no limit to the levels of hierarchical sets of data (or actions) that can be described. We may extend our thinking to describe a set of invoices for a given day.

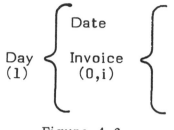

Figure 4.6

Or we can describe a set of days for a given month:

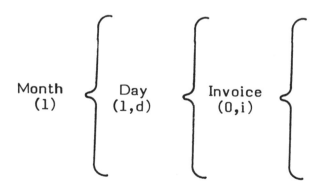

Figure 4.7

With this approach, it is possible to describe an entire system or data base for an organization.

The notation for Warnier/Orr diagrams is easy to master. Essentially, it involves the names of sets (universals), elements (sets or atomic items), and atomic items. Atomic items are items (data or actions) that are not further broken down, at least at this level of description.

As stated previously, set definition in Warnier/Orr diagrams is indicated by a brace. Sequence between elements of a set is implied (top to bottom), alternation (selection) is indicated by the **exclusive or** symbol " ⊕ ," and concurrency is indicated by the **logical or** (and/or) symbol "+."

The repetition of elements is indicated by placing the number of times an element is repeated in parentheses below the name of the element. Allowable possibilities for the number of times include: (0,n), (1,n), (0,12), (1,12), (12), and (0,1).

Warnier/Orr diagrams are ideal for defining structures. Using these diagrams has had a major impact on our perception of the importance of structures in solving complex problems. Warnier/Orr diagrams are, in truth, a language, a logical language. They allow us to describe circumstances in a new and important way. (A detailed description of Warnier/Orr diagrams is contained in Appendix A.)

Mappings

Now that we have tools available for defining sets and structures, let us move on to describing tools that will help us define mappings. Any graphical approach to presenting mappings must have at least the capability of showing three sets or structures: the output set or structure, the input set or structure, and the mapping set or structure.

In Chapter 3 we used a simple form of data flow diagram but we didn't go into it at great length then. Looking at it again, notice, in this case, that we use ellipses to represent data structures, a box to represent the mapping structure between data structures, and arrows to show the flow of the transformation.

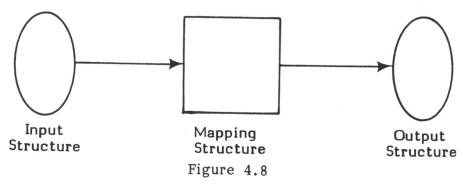

| Input
Structure | Mapping
Structure | Output
Structure |

Figure 4.8

From this point on, we will, in general, drop our reference to sets and refer to structures. Sets are simply the most elementary form of structures.

An alternative means of representing data flow is to show data structures as lines and the mapping structure as an ellipse. The function of the arrows remains the same.

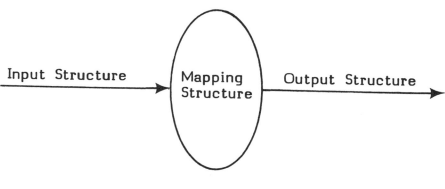

Figure 4.9

We could also show this form of mapping in terms of a special brand of Warnier/Orr diagram called an **assembly line diagram.** An assembly line diagram is a Warnier/Orr diagram in which, by convention, the output structure appears on the left of the brace, the input structure(s) appears on the right of the brace at the top, and the mapping structure appears on the bottom, with the concurrency operator "+" appearing between each of the structures on the right.

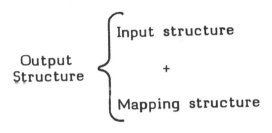

Figure 4.10

In this initial discussion of mappings, we will use the various representations interchangeably. Later, we will adopt the use of the assembly line diagram for most purposes in showing data flow.

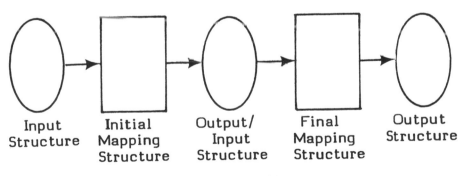

Figure 4.11

The power of data flow diagrams comes not from the singular use of the diagram, but rather from the construction of a series of interconnected mappings, where the output structure of one mapping becomes the input to the next. In this way, one can represent the data flowing through the system.

To develop good systems, we must be able to ensure that our mappings are consistent and correct. The approach we use can be labeled logical analysis. Logical analysis represents the process of applying various logical set operations to decompose or compose new sets or structures.

Let's demonstrate a practical application of logical analysis. Suppose we want to find all those customers who have loan payments more than 90 days delinquent and who owe more than $1,000 (a feedback/control exception report). First, we need to show the relevant universe-- the set of customers of our company.

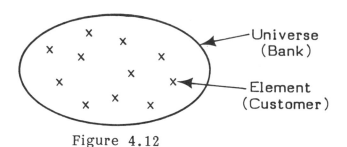

Figure 4.12

Next, we need to subdivide these customers according to their possession (or nonpossession) of certain attributes (having loan payments more than 90 days delinquent and owing more than $1,000).

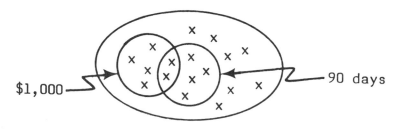

Figure 4.13

The problem can then be described in simple logical terms: sets, subsets, elements, attributes, and relationships. The objective is to list all the customers in the intersection and no one else. We want a mapping.

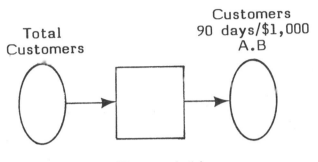

Figure 4.14

This is a simple enough problem, but let's suppose that we have a machine so primitive it can do only one test on one attribute at a time. We would end up with a data flow made up of two mappings: one that maps the total set of customers into a subset of customers (A) who have loans 90 days old, and one that maps that subset into yet another subset (A.B) containing the subset of customers who have loans more than 90 days delinquent and who owe more than $1,000.

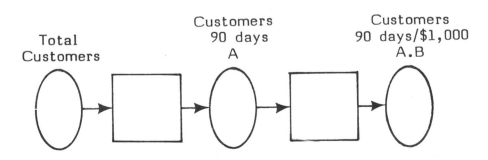

Figure 4.15

The solution mappings are a decomposition of the original problem. Indeed, you could say that the original problem "bounds" the solution. In this case, one could draw a box around the two simple mappings in Figure 4.15 and show that they are a correct solution to the more complex mapping problem posed by Figure 4.14.

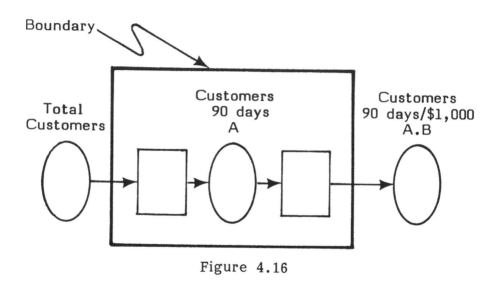

Figure 4.16

Decomposition can be thought of as the breaking down of a single complex mapping into an ordered series of more elementary or simpler mappings. The inputs and outputs of the complex mapping become the boundary of the decomposition. This bounded characteristic makes it possible to think of a mapping and its decomposition as a hierarchy or structure of solutions. And, at a higher level, we can treat the series of mappings as a single mapping and disregard--"hide"--the internal structure. In general, hiding makes it possible to deal with problems from the general to the specific.

All structured methods take advantage of bounded solutions, the process of breaking complex mappings into simple correct ones. In fact, one can actually use a hierarchy diagram where a box (mapping) at one level becomes a boundary at the next lower level.

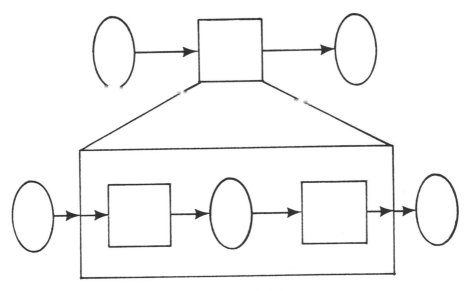

Figure 4.17

Usually, there is more than just one good solution or decomposition. For the problem above (Fig. 4.17), we could also have produced the following primitive solution.

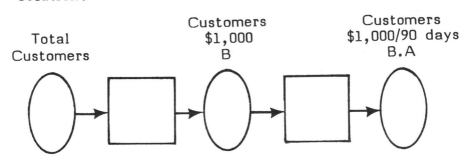

Figure 4.18

In general, if the attributes are logically inde-
pendent, then more than one good solution exists. At
an abstract level, it does not matter which solution we
choose as long as the selection produces the correct
result. In the physical world, on the other hand, the
one we select first may have a major impact on the ef-
ficiency of the solutions. We must, therefore, attempt
to choose optimum solutions, but only after we have
proven their correctness.

Correctness

Any structured approach or tool must first produce
correct answers to real-world problems. Moreover, that
correctness should be easy to verify. In other words,
we need a simple definition of correctness.

A **logically correct procedure** is one in which:

- the **right sets (structures) of actions**
  are done on
- the **right sets (structures) of data** for
- the **right number** of **times** or under
- the **right conditions** in
- the **right order** to produce
- the **right outputs** .

Going back to our original Venn diagram, you will
notice that if we write the actions we want to produce

on the diagram, we have then specified a static view of a **logically correct procedure.** We have a picture that shows what sets of actions are to be applied to specific sets of data.

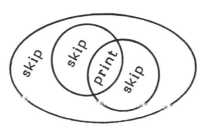

Figure 4.19

Correctness is the hallmark of good systems solutions. By posing the problem as one of mapping structures of actions onto structures of data, correctness becomes a workable concept.

Over the years, a number of different techniques have been developed to show the mapping of simple sets or subsets of actions onto simple sets or subsets of data. Three forms of logic diagrams (Karnaugh map, decision table, flowchart) are shown below.

	90	$\overline{90}$
1,000	print	skip
$\overline{1,000}$	skip	skip

Figure 4.20. Karnaugh Map

Rules

90 days	T	T	F	F
1,000	T	F	T	F
Print	x			
Skip		x	x	x

Figure 4.21.   Decision Table

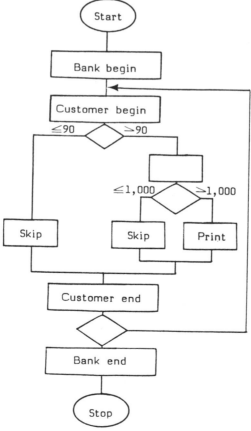

Figure 4.22.   Flowchart

Each view describes the application of the rule of correctness to sets and subsets of data. Unfortunately, these approaches fall short when we attempt to apply them to "structures" of data and actions. For this reason, even though many schemes for subdividing problems and for applying the various logical tools to the pieces have been developed, these techniques have had limited application.

And if this applies to simple mappings involved with structures, it becomes even more apparent when we attempt to define a dynamic situation. Let's look again at our bank problem. We can show a dynamic view of the same situation with a "bubble chart."

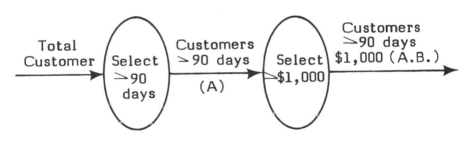

Figure 4.23

This diagram captures the essence of what we are trying to accomplish: to create one set of data from another. Moreover, if we wish to develop a more complex set of mappings from it, we can do so.

Data Flow Design

Data flow diagrams are powerful tools for representing mappings. In fact, data flow diagrams in one form or another have been used for decades to design and document systems. The systems flowchart, for instance, is a type of data flow diagram.

To the professional in the field, the use of this kind of tool is assumed; but in the majority of training programs, instruction in such tools has been negligible. Part of this deplorable lack of training in the basics stems from the fact that the traditional data flow diagrams have serious limitations. The more powerful the tool, the more apt it is to be abused. In the hands of a well-trained analyst or designer, a data flow diagram can be a formidable weapon; in less skilled hands, it can present problems.

Recently, however, Constantine, Yourdon, Ganes and Sarson, and DeMarco have popularized the data flow diagram as a means of defining systems. For the most part, the diagrams they employ use bubbles or boxes to represent mappings, and lines to represent data sets or structures.

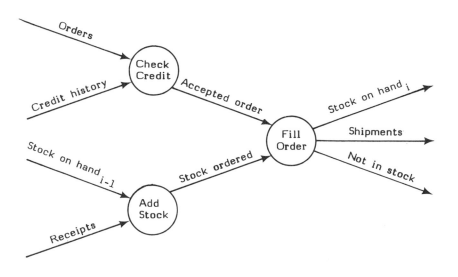

Figure 4.24

One of the difficulties with the standard type of data flow diagram is that it can become complicated to work with. This normally occurs when various mappings are shown to produce multiple outputs and those outputs go to many places. The result is a network of inter-secting lines.

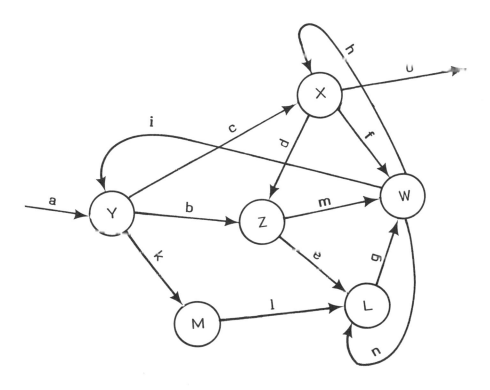

Figure 4.25

Such a picture can be an absolutely accurate repre-sentation of the flow of information within an organi-zation. But it is of little help in developing a cor-rect, efficient, logical solution that can be mapped into a workable system.

The way around this problem is to insist that data

flow diagrams be based on "functional mappings." Larry
Constantine, one of the leading theoreticians of the
data flow school of design, has developed a scale of
mappings, the highest of which is functional. *A functional mapping has only one output.* The necessity of
imposing the single-output rule can be seen by analyzing a portion of the data flow diagram (Fig. 4.25).

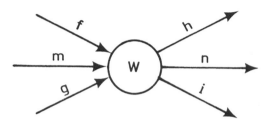

Figure 4.26

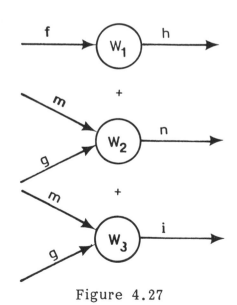

Figure 4.27

On the face of it, one has no way of knowing any of
the dependencies of outputs on inputs. We are left to

assume that everything depends on everything else. If
we decompose "W" into a series of concurrent mappings,
however, we may find that this is not the case.

Once we have broken our original mapping down into
its components, we can see that we are dealing with a
problem that contains apples and oranges (Fig. 4.27).
For example, W1 has no business being lumped with the
others. W2 and W3 do have a lot in common, and in
physical design, we may want to "package" them together
into a single program or module.

Our objective, then, is to retain the power of the
data flow diagram but to avoid the problems "bubble
chart" type diagrams may pose. To reach our objective,
we use the assembly line diagram. The assembly line
(or functional flow diagram) corresponding to Figure
4.25 is shown in Figure 4.28.

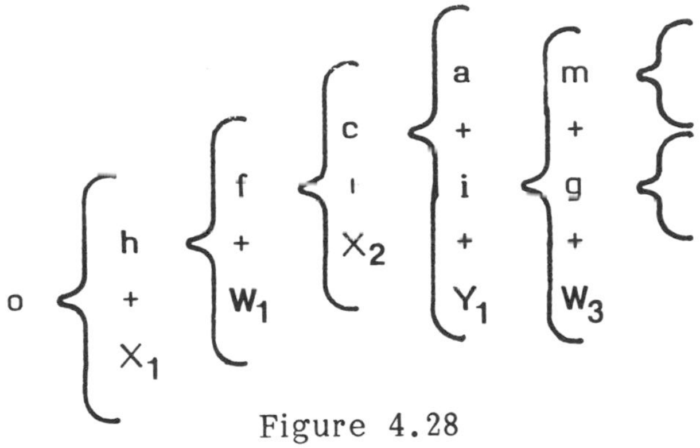

Figure 4.28

In an assembly line diagram, an output set or structure "comes from" some number of input sets and from
a mapping or process. In turn, each of these inputs,
except for the ultimate inputs that appear as leaves
of the hierarchy on the right, becomes an output that
"comes from" yet another set of inputs and a mapping.

The power gained through the use of the assembly

line diagram is considerable, for with such a logical picture, we can trace the "genealogy" of each output. Information must come from somewhere, and the assembly line diagram can be used to trace each output back to all the inputs that went into its generation.

A certain amount of overhead is incurred by using the assembly line diagram. It is the result of requiring each mapping to produce only a single output. This seems needlessly redundant to anyone accustomed to using other forms of data flow diagrams. The redundancy is important, however, and well worth the price.

If one currently has a process or program with multiple outputs, it can be converted by a simple functional decomposition.

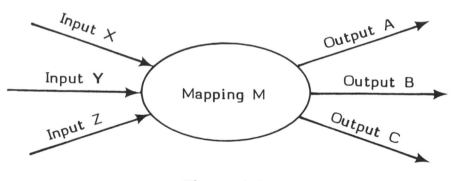

Figure 4.29

That figure becomes the assembly line diagram that appears in Figure 4.30.

In defining structured requirements, assembly line diagrams are used to develop the functional flow or main line of the system. In truth, the functional flow is nothing more or less than a data flow diagram showing the important mappings within the system. Thinking of such a flow as an "assembly line" has aided both in understanding and teaching this approach.

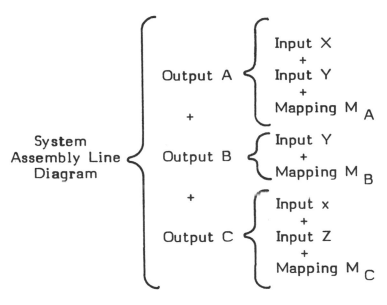

Figure 4.30

In case you're wondering how assembly line diagrams got their name, I'll explain. After teaching output-oriented design for some time, I needed some means of displaying this approach in a graphical manner. I tried various methods until I finally stumbled across the idea of representing a data flow as an assembly line, and it then occurred to me to teach design as a process of "disassembly."

Let us consider the design of a real assembly line, e.g., an assembly line to produce automobiles. How would you go about it? This is actually a complicated process involving thousands of individual steps, each of which has to be done in exactly the right order. An output-oriented design approach starts with the final output and disassembles it a little at a time, taking off one thing at a time from the outside, making subdivisions at some basic interface.

First, we might imagine a finished car.

Figure 4.31

What would be the first thing to take off? Well, in my neighborhood it was always assumed that the hubcaps were the easiest things to remove.

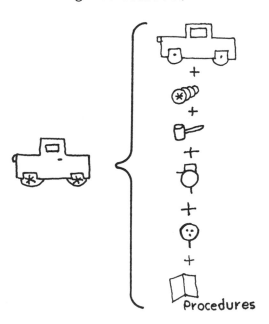

Figure 4.32

Our initial assembly line diagram might show the following elements:

- a fully assembled car but **without** hubcaps (input),
- a set of hubcaps (input),
- a rubber hammer (tool),
- an employee to put on the hubcaps (agent, mechanism),
- one or more people to monitor the process (control), and
- a procedure for putting on hubcaps (mapping).

At this stage, instead of one problem, we have two major subproblems: to assemble a car without hubcaps, and to build a set of hubcaps. This disassembly process can be carried on recursively, subassembly by subassembly, asking at each stage, "Do we make this assembly (subassembly) or do we buy it?" When there are no more pieces to disassemble, we are at a stopping point.

Although this is a somewhat far-fetched application, the principle is fundamental to all design and planning. The only foolproof method for designing anything involves some form of starting with the goal (output) and working backward.

Since we began teaching disassembly (working backward) as a method of design some years ago, numerous examples of its application in other fields have been brought to our attention. For example, one student said that he was frankly skeptical about using this approach in data processing until he remembered that this was exactly the way engineers in his organization had set up an actual assembly line. In this instance, the engineers literally placed a test model on the end

of the assembly line and took pieces off. As they took
them off, they noted exactly what they had done; when
they were finished, they simply reversed the order to
create the real assembly line.

This approach was used in a project undertaken by
Bob Langston and Langston, Kitch and Associates, Inc.,
several years ago to develop large-scale PERT/CPM net-
works for an airplane manufacturer. The task involved
building a network that tied together the tens of thousands of individual activities required to design and
build a multi-million-dollar airplane for the Department of Defense. Starting at the beginning proved
fruitless. Finally, it was concluded that the only
practical approach was to write down the last activity
(in this case "fly away") and work backward from there.
This proved very successful.

Some people have argued that this approach will not
work in data processing because information systems
have a great many different outputs, and to develop a
separate assembly line for each would be unthinkably
inefficient. Despite this argument, we have found that
the disassembly process does work in designing information systems, and it works well. Although we do develop separate, logical assembly line processes for each
major output, or group of similar outputs, we find that
as we work backward, certain common, intermediate subassembly lines occur at just the points where good systems have traditionally broken into modules. Moreover,
the separate assembly lines provide a functional picture that can be used in physical design to package the
system into efficient modular units.

In requirements definition, Warnier/Orr diagrams are
used to structure outputs and to show the cycles involved in the production of the outputs. Assembly line

diagrams are used to delineate the flow of information in an application, and thereby to identify the outputs that have to be defined. That there are two different uses for Warnier/Orr diagrams, one to show hierarchical structure, and one to show data flow, is not surprising. All the different methods of structured analysis and design (SADT®, HIPO, Constantine-Yourdon, etc.) have similar breakdowns.

Constantine was perhaps the first to recognize that the data flow view should come before the hierarchical control view. He also recognized the necessity of having a systematic means of mapping the data flow into the hierarchical control view. The methods he proposed were primitive, but the idea was basically correct, and in fact, we have developed just such a method through an analysis of the cycles of the system.

Entity Diagrams

In the early days of Structured Systems Design, some users had difficulty getting started with the approach. Our basic rule, "define the outputs," was useful, but it often did not go far enough. Various systems analysts asked, "What if the user doesn't know what the outputs are?" Our second basic answer, "Work on it!" didn't seem to be satisfactory.

About this time our firm was involved in designing an information system for a government agency. In addition to asking how to get started, the client also came up with some other questions: "How do you know when you have all the outputs?" and "How do you know if you've got the right outputs?"

The leader on this project, Pete Kitch, put this

point very graphically. "What I would really like," he
said, "is an output tree. That way I would know where
to go to find all the outputs and when the last one was
pulled off, I would know I was done."

The attempt to answer the questions:

- How do you get started?
- How do you know when you have
  all the outputs?
- How do you know if you've got
  the right outputs?

led us to become more specific with our requirements
definition methodology. Out of this work evolved the
entity diagram.

Entity diagrams came into being to help get the project started. During one of the initial meetings some-
body suggested, "Let's define who does what." So, for
each entity involved, a circle (ellipsis) was drawn,
and the name of the entity placed inside it. Then the
data that passed between the various entities was defined, with arrows showing the direction of the information flow. With the various entities clearly delineated in this diagram, the project team was able to get
off dead center with the user. Since that time, we
have used the entity diagram to help us define the objectives and, subsequently, the functional flow assembly line for the system.

Chapter 6 will describe exactly how to use entity
diagrams, assembly line diagrams, logical data layouts,
and Warnier/Orr diagrams for requirements definition.
Here we simply want to show you what goes into an entity diagram.

We will use a payroll system to illustrate the de-

velopment of an entity diagram. What entities (external) are involved? First, there is the company; then
there are the employees; the federal and state governments; the insurance companies (health and life); the
unions; the united fund, etc. There may be others,
such as credit unions and political action committees;
depending upon the situation these would also be placed
on the diagram.

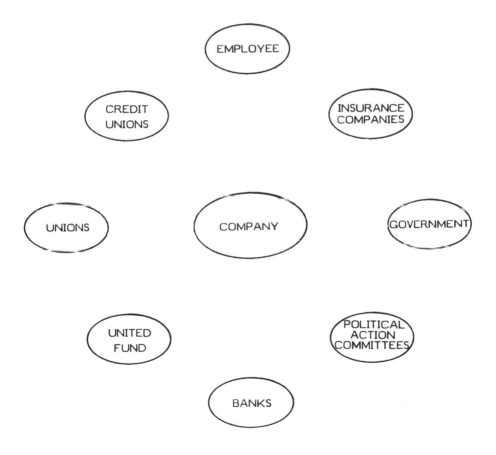

Figure 4.33. Defining the Entities

Figure 4.33 defines those entities. Once they are defined, the next step is to delineate the transactions between entities, and those are shown in Figure 4.34.

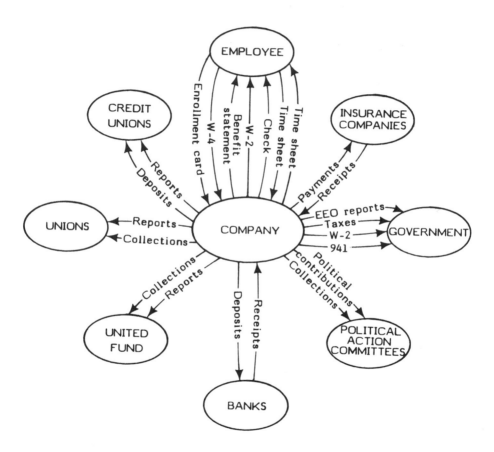

Figure 4.34. Entity Diagram - Payroll System

The entity diagram is not usually pretty in the beginning, since the basic purpose is to provide the analyst or user with a means of getting "unstuck." Gregory Bateson talks about two stages of thinking, a "fuzzy thinking period" and a "clear thinking period." An entity diagram is a fuzzy thinking device.

The process of systems work is subtle. At one end of the spectrum is the user's environment, which is high on context and low on precision; at the other end is the data processing system, which is low on context (everything is defined) and high on precision.

Many fundamental data processing problems stem from the fact that both users and analysts assume a great deal. The user assumes that the analyst understands the meaning of his dialect of English (organizational jargon). And the analyst expects the user to think in precise fundamental terms as he does.

Extended systems work, even without the use of many of the structured tools and techniques, leaves its mark upon the thinking of the analyst. He begins to perceive the world through a series of abstractions and in ways that are foreign to ordinary thinking. As a consequence of learning the systems trade, it becomes somewhat difficult to remember how things used to appear when one knew little or nothing about inputs, outputs, files, and programs.

The user often works in a very narrow world, taken up with day-to-day problems. He does not see a system at all, but rather a series of recurring happenings. He may know intuitively what is wrong, but he has neither the faintest idea how to express his problems nor a method for curing them.

The entity diagram, then, provides a way to define the context in which systems improvement can take place. It is a tool for helping user and analyst communicate with each other to set the scope of the system. As we shall see, the entity diagram has a language of its own.

In retrospect, it is clear to see that the entity diagram is simply another way of representing a set of mappings. Clearly, this diagram has more than a family resemblance to a data flow diagram. The "entities" can be considered transforms where the arrows represent (or can represent) information sets.

But the entity diagram is even more like a state diagram. Indeed, as we worked with them, we discovered that entity diagrams "model" many significant factors associated with systems definition.

The segment of an entity diagram in Figure 4.35 shows a number of major factors (entities and transactions), but one can deduce even more. For example, at the origin and termination of each transaction (arrow), there must be **events**. The "sending" and "receiving" of an order is important to the system. Thus events important to the system can be "read" off the diagram in a systematic manner.

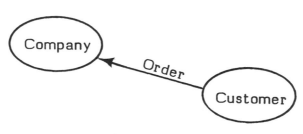

Figure 4.35

The entity diagram proves again that one good picture is worth a thousand words. But it is only a starting point for the requirements process. In the last stages of requirements definition, the stress is on the precise definition of the outputs. For this purpose, logical data layouts, mock-ups or sample reports, and Warnier/Orr diagrams are used. As a by-product, a data dictionary is produced.

Logical Data Layouts (Models)

In defining outputs, we need to decide what data is required and how it will be presented. The **logical data layout** or **logical data model** is a tool for doing this. Like the Warnier/Orr diagram, the logical data layout first appeared in Warnier's early work, but he had not given this particular tool a name. We subsequently called it the logical data layout. (Warnier now calls it the "model of the requested output" or, simply, the "model.")

Warnier recognized the need for a means to present a logical picture of the data required. He knew that physical data layouts (large blue sheets with **x's** and **9's** on them) just did not do the job. As with so many tools and techniques we have borrowed from Warnier, the logical data layouts turned out to be very useful.

The logical layouts consist of three elements:

(1)   a boundary;
(2)   "buckets" for atomic elements; and
(3)   the names of atomic elements.

We start with a blank page (or screen), and that's where the boundary comes in.

Output/Input
Page or Screen

Figure 4.36

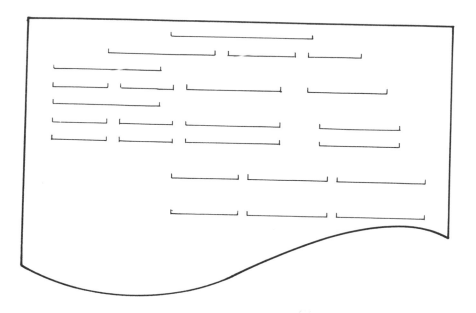

Figure 4.37

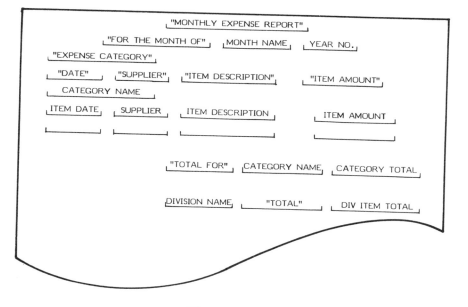

Figure 4.38

Within this boundary we place brackets for each atomic element, variable or literal, (Fig. 4.37). In Figure 4.38, we place the names of each atomic element (or literal) above the brackets.

Sample Outputs

Logical data layouts do not, however, take the place of **sample reports** or **simulated screens.** Most users have serious difficulties dealing with abstract definition. Thus, in structured requirements definition, we go to great lengths to make sure that the user knows exactly what he will be getting from the system and why. This requires extra effort in the requirements definition, but it is amply repaid in the elimination of confusion. We develop sample outputs for the same reason that architects make preliminary sketches and build scale models--to make certain the buyer can predict precisely what he will receive (see Fig. 4.39).

Data Dictionaries

On any project, there is a need to have a consistent set of names to use when referring to various pieces of information. For example, we need names of data elements, records, programs, procedures, etc. Over the years, dictionaries of various kinds have been developed for just this purpose.

With the advent of computerized data-base management systems has come the recognition that it is mandatory to have a consistent naming scheme. If data items are not referenced with the same names or if they are given

```
                        MONTHLY EXPENSE REPORT
                      FOR THE MONTH OF MARCH 1979
      EXPENSE CATEGORY
      DATE      SUPPLIER           ITEM DESCRIPTION          ITEM AMOUNT

      PRINTING

      03-10-79  G.H. PRINTING      FUTURES                      450.00
      03-15-79  G.H. PRINTING      PROMOTIONAL MATERIAL         550.00
      03-25-79  XX BLUEPRINTING    SSD NOTES                    100.00

                                   TOTAL FOR PRINTING         1,100.00

         .          .                  .                          .
         .          .                  .                          .
         .          .                  .                          .

      TRAVEL EXPENSES

      03-08-79  K.ORR              CARSON CITY, NEVADA          525.00
      03-15-79  P.KITCH            LANSING, MICHIGAN            375.26
      03-25-79  P.KITCH            CINCINNATI, OHIO             400.18

                                   TOTAL FOR TRAVEL EXPENSES  1,300.44

                                        .
                                        .
                                        .

                      ADVANCED SYSTEMS DIVISION TOTAL       45,907.95
```

Figure 4.39.   Sample Report

different names by different users, then the data-base management process becomes impossible. The answer to this problem has been to develop data dictionaries.

Data dictionaries are simply a reference point for information about the data in the system. In our data dictionaries, we maintain information about the following categories of data:

- **Sets (Logical Files)** - name, attributes, synonyms, where used
- **Elements (Logical Records)** - name, attributes, computation/decision rules, synonyms, where used

- **Atomic Elements (Logical Data Fields)** - name, attributes, computation/decision rules, synonyms, where used
- **Relationships (Logical Structures)** - name, attributes, structure, synonyms, where used

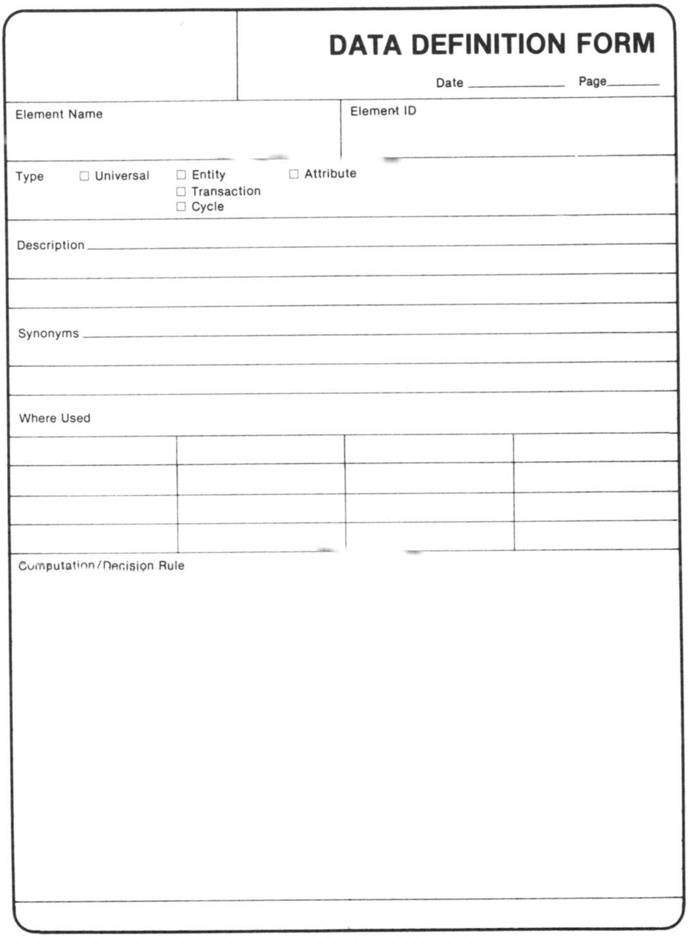

DATA DEFINITION FORM

Date _____ Page_____

Element Name

Element ID

Type ☐ Universal ☐ Entity ☐ Attribute
                    ☐ Transaction
                    ☐ Cycle

Description _____

Synonyms _____

Where Used

Computation/Decision Rule

Figure 4.40

There is one important difference between the way data dictionaries are created in structured systems development vs. other methods. Basically, the information goes into the data dictionary a little at a time, starting with the definition of the outputs. As a result, the data dictionary is produced as a by-product of the systems development process. In the early stages, most of the data is at a logical level. Physical data, e.g., field length, coding, etc., is entered later when it is precisely defined.

ID	NAME	DESCRIPTION	TYPE	COMPUTATION/DECISION RULES	SYNONYMS	WHERE USED
		DATA ELEMENT DESCRIPTION **BY DATA ELEMENT NAME**				
DE-5	AGENT.NAME	NAME OF CUSTOMER'S AGENT				R-01
DE-11	CATEGORY.NAME	NAME OF A MAJOR GROUP OF PURCHASES				R-02,R-03
DE-15	CATEGORY.TOTAL	TOTAL AMOUNT OF ITEM AMOUNTS FOR A CATEGORY FOR THE CURRENT MONTH	C	ITEM.AMOUNT		R-02
DE-17	CURR.MONTH.AMT	TOTAL AMOUNT OF ITEM AMOUNTS FOR A CATEGORY FOR THE CURRENT MONTH, CURRENT YEAR	C	ITEM.AMOUNT		R-02
DE-18	CURR.YEAR.AMT	TOTAL AMOUNT OF YEAR TO DATE ITEM AMOUNTS FOR A CATEGORY FOR THE CURRENT YEAR	C	ITEM.AMOUNT		R-03
DE-2	CUST.TOTAL	TOTAL AMOUNT OF INVOICE FOR CUSTOMER FOR CURRENT MONTH	C	INVOICE.AMOUNT		R-01
DE-3	CUSTOMER.NAME	NAME OF CUSTOMER				R-01
DE-21	DIV.CURR.MONTH.AMT	TOTAL AMOUNT OF ITEM AMOUNTS AMOUNTS FOR A DIVISION FOR THE CURRENT MONTH, CURRENT YEAR	C	ITEM.AMOUNT		R-03
DE-22	DIV.CURR.YEAR.AMT	TOTAL AMOUNT OF YEAR TO DATE ITEM AMOUNTS FOR A DIVISION FOR THE CURRENT YEAR	C	ITEM.AMOUNT		R-03
DE-1	DIV.INV.TOTAL	TOTAL AMOUNT OF INVOICE SALES FOR DIVISION FOR MONTH	C	INVOICE.AMOUNT		R-01
DE-16	DIV.ITEM.TOTAL	TOTAL AMOUNT OF ITEM AMOUNTS FOR A CATEGORY FOR A DIVISION FOR THE CURRENT MONTH	C	ITEM.AMOUNT		R-02
DE-23	DIV.PREV.MONTH.AMT	TOTAL AMOUNT OF ITEM AMOUNTS FOR A DIVISION FOR THE CURRENT MONTH, LAST YEAR	C	ITEM.AMOUNT		R-03
DE-24	DIV.PREV.YEAR.AMT	TOTAL AMOUNT OF YEAR TO DATE ITEM AMOUNTS FOR A DIVISION UP TO THE CURRENT MONTH, LAST YEAR	C	ITEM.AMOUNT		R-03
DE-25	DIVISION.NAME	NAME OF DIVISION OF COMPANY				R-01,R-02,R-03
DE-7	INVOICE.AMT	AMOUNT OF SALE OF INVOICE				R-01
DE-8	INVOICE.DATE	DATE OF THE PREPARATION OF INVOICE				R-01

Figure 4.41

In general, we do not promote promiscuous collection of information for the data dictionary. The data dictionary should contain only items that are used. Only elements that appear on outputs or elements used in computing elements that appear on outputs need to be defined. If there is a need to keep note of various terms or expressions that may be important at a future date, that information should be placed in a glossary, which is strictly free form.

The form in Figure 4.40 is used to document data for the data dictionary. Figure 4.41 shows outputs produced from such a data dictionary.

Naming Names - The "Of" Language

There are many stories about the significance of names. One such story has to do with Confucius. Once, when the emperor was thinking of making Confucius prime minister, he called the philosopher to the palace. The emperor asked, "What would your first act be upon becoming prime minister?" Confucius replied, "I would issue a proclamation requiring that all things be called by their right names." Needless to say, Confucius was not made prime minister.

What is important is not just the names we give things, but the approach we use for naming. It is critical to have a taxonomy, a system of naming. This is particularly true in large systems, where there may be thousands of different names.

Some years ago, a project team at IBM was assigned the task of building a data dictionary for the entire corporation. One of the by-products of that activity was something called the "of" language. "Of" language is a method of assigning data names from the general to the specific, using certain words as connectors. Like most powerful ideas, the "of" language is extremely simple (see the examples in Figure 4.42). As a result

of assigning data definitions in this manner, it becomes clear that most atomic data elements fall into a
limited set of categories: names, identification
codes, amounts, totals or computed values, and times.

NORMAL DATA NAME	"OF" LANGUAGE
CUSTOMER-NAME -ABBREVIATED	NAME (of) CUSTOMER (which is) ABBREVIATED
SHIP-DATE	DATE (of) SHIPPING (of an) ORDER (assigned by) MARKETING
EMPLOYEE-NO	CODE (to) IDENTIFY EMPLOYEE (assigned by) PERSONNEL

Figure 4.42

Having the right names for things is critical. Many
systems analysts have gotten themselves and their
organizations into serious problems because no one
thought to question what was meant by some innocuous
data field such as SHIP-DATE. (I have worked on applications where there were as many as three different
meanings for SHIP-DATE, depending on whom you were
talking to.)

Invariably, it is the things everyone knows that get
us into trouble. The application of a systematic data
definition can eliminate the causes of many problems in
communication.

Theory/5

****

Form follows function.

Frank Lloyd Wright

In Chapters 3 and 4 we considered the fundamental concepts and tools that underlie structured requirements. In this chapter, we will use these concepts and tools to explain the theory associated with structured systems development.

As a way of conceptualizing the systems environment, one can think of an **organization** (or primary systems user) and the **real world** (or environment) as entities interacting with one another. (In point of fact, the organization is part of the real world, but considering it as separate will serve as a useful abstraction.)

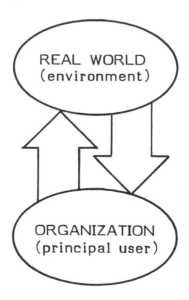

Figure 5.1

In this case, the arrows that represent transactions are enlarged to indicate the set of all interactions between the organization and the world. In a sense, the total system and the organization can be thought of as identical. There is only one organization and, ultimately, only one system.

Viewed from this level of abstraction, we can see that the organization interacts with the real world: (1) by producing products (services) and actions, and (2) by accepting various inputs.

In general, the real world sends the organization far more information than the organization can profitably use. This is true not only of organizations, but of all organisms. A human being processes only a very small portion of the spectrum of data that impinges upon the senses. For example, above and below the range of light waves visible to human beings, there are enormous ranges of wavelengths that are imperceptible (unperceived). This narrow selectivity is not accidental--biology sees to that. Organisms sense only those things that are necessary to the survival of the species. Using our model, this means that an organization, like an organism, will only be interested in (or aware of) a relatively small subset of its total interactions with the environment (see Fig. 5.2).

The products and services of an organization will define those portions of the real world that the organization considers important. If the organization builds and sells automobiles, then it will concern itself with things automotive--automotive technology, automotive regulations, and automotive customers. If the organization builds supertankers, however, it will concern itself with a much different part of the real world.

Most organizations are exceedingly aware of the areas and entities of the real world that affect the

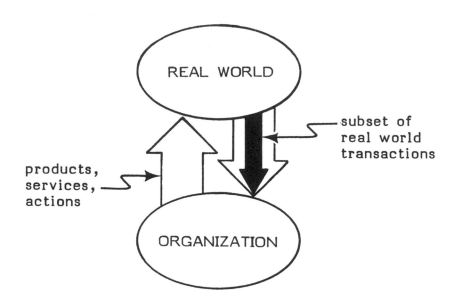

Figure 5.2

organization's well being, at least in the short run.
If the organization chooses to restrict its inputs, the
end result is that each organization deals with something less (in terms of data) than the "everything" the
real world sends its way. This means that the organization only deals with a selected subset of the entities and transactions that make up the environment (see
Fig. 5.3).

Organizations are not only sensitive to different
subsets of the real world, their sampling rate varies
as well, depending upon the rate of change of entities
that most concern them. For example, suppose a petro-
chemical manufacturing organization has three divisions
specializing in different products. One division manufactures bulletproof plastics for windows, another manufactures soles for shoes, and a third manufactures
fertilizer for crops.

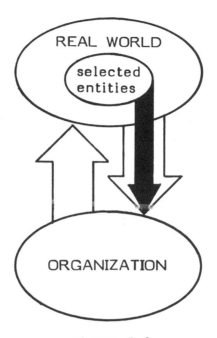

Figure 5.3

In each division research scientists are studying the effect of new materials on products and need to **see** what happens when the new materials are incorporated into the products. To visualize these effects, the scientists take motion pictures of experiments at an appropriate speed, and play back the results at rates that are compatible with human processing.

In all cases, the data is displayed at normal projection speed, just fast enough so the person watching does not notice flicker. But for a researcher to be able to observe changes that are critical to that particular work, the rate at which the input is collected will be vastly different. In the case of the scientist developing bulletproof plastic, the rate of input might run 1/1000 of a second to catch the bullet or projectile at the moment of impact.

Vice versa, if the scientist studying shoe soles is interested in the ability of the sole to withstand wear, input might be collected very slowly, and played back at a high rate of speed. Similarly, the scientist studying the effect of a new fertilizer might capture his input at a rate measured, perhaps, in terms of hours or days. (A scientist studying glaciers might only sample on a monthly or yearly basis.)

Different organizations are sensitive to differing aspects of the real world based on what factors they are trying to control. This last point is so critical that it cannot be overemphasized. Organizations, organisms, and systems alike are goal oriented. They all attempt to maximize or minimize something. To do this they use feedback to monitor the disturbances in the real world. If a piece of information is directly related to the goals they are trying to achieve, then they will be sensitive to it; otherwise they will not.

For instance, a manager of an organization whose primary goal is growth is likely to be sensitive to changes in sales and in expanding sales potential, but may not be as sensitive to changes in profitability or costs. On the other hand, a manager in a highly regulated business would be sensitive to the ratio of profits to capital expenditure, but might not be as interested in market growth, and so on.

The lesson to be learned from the examples above is that *different organizations are sensitive to different entities and transactions in the real world, as a result of their products, services, actions, and rates of change, and goals.* An organism is sensitive to information that it uses. Therefore, when we study the requirements of an information system, we should be careful to take advantage of that fact.

A systems scientist who is developing the requirements for a system must be aware of the nature of the
business, the entities of the world that must be dealt
with, the rates of change, and the organizational goals
and objectives. Otherwise, the resulting system may be
out of sync with the organization.

An organization deals only with a subset of the entities and transactions in the real world, whereas an
information systems application (application for short)
deals only with a subset of that subset of the entities
of interest to the organization.

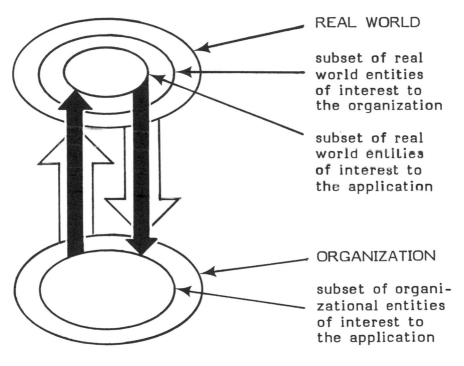

REAL WORLD

subset of real
world entities
of interest to
the organization

subset of real
world entities
of interest to
the application

ORGANIZATION

subset of organi-
zational entities
of interest to
the application

Figure 5.4

An **application** (or information system) is a collection of related entities and transactions that make

up an essential, feasible, functional, workable, correct, and operational model within the organization.
These terms have been chosen carefully to indicate the
most important characteristic of each phase of the systems life cycle.

PHASE	SUBPHASE	CHARACTERISTIC
PLANNING	LOGICAL	ESSENTIAL
	PHYSICAL	FEASIBLE
REQUIREMENTS DEFINITION	LOGICAL	FUNCTIONAL
	PHYSICAL	WORKABLE
DESIGN	LOGICAL	CORRECT
	PHYSICAL	OPERATIONAL

Essential

An **essential application** is necessary to the operation of the organization, solves major problems or
makes possible the exploitation of major opportunities,
and is a major priority from an organizational standpoint.

The systems planning process is primarily focused
upon determining which of the wide variety of applications that could be developed are actually carried forward to the requirements definition phase. "Is this
application really necessary?" stands out as the key

question in logical systems planning.

In structured systems development, essentiality is defined in terms of identifying: problems (opportunities), user(s), uses, and context of the application.

Feasible

A **feasible application** is judged to be economically, technically, and organizationally possible given the organizational environment. While a great many applications have a nice ring to them, when analyzed they prove to be deficient in one or more of the categories of possibility.

Economic feasibility is reasonably well understood. In general, most organizations are careful to judge the economic risks involved in developing computer applications, even though most data processing organizations have a poor record of estimating development costs. In nearly all cases, there is some form of estimating and approval approach involved in initiating projects.

Even though organizations try to estimate the cost of development and operation of their applications, however, not enough attention is ordinarily given to the cost (or economic risk) should the system fail--or not perform according to initial projections. A great many ill-conceived applications are undertaken without any real attention to their exposure.

The situation is even more difficult in the case of technical and organizational feasibility. There is often a great difference between the ideal and the real world when it comes to advanced data processing applications. Researchers have speculated that the first application of a new technology in an organization, even where the technology is worked out, costs two to four times as much as does an application using a "known" technology. This is generally referred to as the "learning curve."

The case is even more expensive (and risky) where the technology has not been thoroughly shaken down. In the 1960s, many organizations learned, to their dismay, that what looked good on paper didn't quite work in practice.

In structured systems development, feasibility is defined in terms of economic, technical, and organizational benefit (risk).

Functional

A **functional application** does what it has to do, and only that. A functional system is one that works well and operates in a consistent manner for a number of different users, managers, and operators.

Over time, we see a progression of users and requirements, and we would like our systems to continue to operate, even in the face of these changes. Indeed, we would like our systems to produce the correct outputs from any normal inputs, and we would like them to do so under the full range of normal conditions.

Functional, in its fullest sense, also means ignoring the internal organization structure and procedures. Truly functional systems span ill-defined organizational, as well as systems, boundaries.

In structured systems development, functions are defined in terms of context, functional flow, functional steps (subsystems), decision control feedback, scope, outputs, and cycles.

Workable

A **workable application** can be operated by organizational personnel, and supported economically, technically and operationally.

The notion of workability is related strongly to the constraints imposed by reality. A functional system is defined by what has to be done in order to support some organizational goal, without concern for response times, volumes, human factors engineering, and a myriad of physical constraints. Workability, on the other hand, brings these factors into play.

It is of little interest to produce a system that cannot, or will not, be used within the organization. Therefore, there is a concentration in requirements definition upon applying physical constraints to the functional requirements.

In structured systems development, workability is defined in terms of constraints, alternative solutions, risks/benefits, and recommended solutions.

Correct

A **correct application** produces the desired results at the required times, according to the required calculation and decision rules. The first and foremost technical issue in design is whether the application produces the right answers all the time. In structured systems development, we have placed a strong emphasis upon defining the logical correctness of our systems, based on the results and constraints determined in requirements definition.

Over the last decade, we have come to understand the

vital importance of logical design being completed before the physical features are introduced. The test of logical design is at all times the ability to prove correctness.

In structured systems development, correctness is defined in terms of the logical specifications (blueprints) of the application.

Operational

An **operational application** is efficient to operate, robust, flexible, and maintainable. Efficient means capable of executing a given logical design within the existing physical constraints and using a minimum of resources. Robust means capable of detecting, limiting loss, and recovering from normal operating errors. Flexible means capable of being taken apart and put back together. Finally, maintainable means capable of being changed easily.

In the same way that certain functional systems are not truly workable, so too, some correct systems are not operational. The nature of physical design is to incorporate into the systems those features that make the system easy to operate, efficient to run, capable of recovery, etc. In the early days of structured design, one of the common errors was to equate logical and physical design. We learned the hard way that, even with the best logical design in the world, there is a definite need for a logical procedure of physical design.

In structured systems development, operationality is defined in terms of a set of physical specifications (blueprints).

Any methodology for building systems must deal with all the concepts defined above. In structured requirements definition we are concerned primarily with that piece of the methodology that fits between planning and design. We can assume, since we are within a structured life cycle, that an application has been defined that is both essential and feasible. It is the purpose of requirements definition to further refine our understanding of the application so it is functional and workable as well.

The process that is described in Chapter 6, Procedures, falls within this framework. There is a time and a place for everything, and logical requirements definition is a time for defining the context, main line, functional steps, decision control system, outputs, and cycles. Physical requirements definition is a time for defining constraints, alternative solutions, risks/benefits, and recommended solutions.

Procedures/6

****

A human procedure is as important to the
system as a machine procedure. People have
radically different instruction sets than
machines, including an operation called
"use your common sense," but they
have instruction sets just the same.

Harlan Mills
IBM Systems Journal

In critical human activities, such as performing open heart surgery or flying an airplane, where the slightest mistake can be fatal, the precise steps involved are nearly always committed to writing and followed religiously. The result is a procedure--a set of steps to accomplish a task.

Procedures are a means by which theoretical knowledge, research, and experience are put into practice. If you want to put a policy or an idea into practice, you can enact it into law, or teach it in school, or develop a step-by-step procedure, with checklists to make sure the procedure is followed. Of these approaches, creating a procedure is perhaps the most effective.

In many respects, civilization is the result of translating knowledge into procedures. Civilized life is predicated upon thousands, even millions, of procedures. There are procedures for erecting buildings, digging tunnels, removing teeth, installing telephones, and so on. A cookbook contains hundreds of procedures. A musical score is a precise procedure. Doctors and lawyers spend the better part of their careers learning and perfecting procedures. In general, the more taxing and complex the activity, the more emphasis is placed upon procedure.

Surely, defining the requirements of an advanced information system is complex enough to warrant the development of a step-by-step approach. That is precisely what we have attempted to do with all the phases of structured systems development. This chapter, then, is concerned with defining and showing examples for the procedure called structured requirements definition.

The procedure presented here is only a guideline. It is intended to be employed by experienced systems scientists, or beginning systems scientists under the guidance of superiors. The procedure is not a substitute for hard work, imagination, or insight. It does provide, however, a tested method for aiding information systems users in defining their requirements in a systematic fashion. Those requirements can then be used to build a correct structured system.

We have chosen to use our own tools to describe the procedure involved in requirements definition, namely Warnier/Orr diagrams. We don't have anything against natural language; it is simply that we have found that this manner of describing a procedure is much simpler to explain, and communicates better than any other scheme we have yet stumbled upon.

Clearly, this procedural description can be translated into other logical languages; however, as with any translation, something may get lost. The same is true of the other tools we employ. It is clearly possible to use the procedure outlined in this chapter with other forms of structured and unstructured documentation. But since the procedure and tools have evolved together and are the result of a theoretical basis, the mixing of other techniques may not work as well.

In general, procedures are a function of the things they produce. Requirements definition produces a product used to design and build information systems. From Chapter 5 we know that requirements definition should produce an application, both functional and workable, from a planning document that identifies an application as essential and feasible.

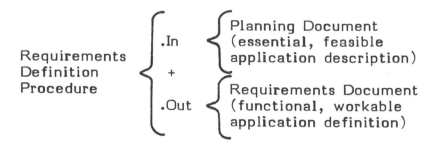

To be specific, the requirements document of a functional, workable system must include:

- a definition of the principal outputs
  (layout, sample, structure, decision
  and calculation rules, volume, frequency
  and response time)
- a dictionary of data definitions
- a definition of the assumptions and
  constraints involved in the system
- a definition of the risks and benefits
  of various approaches

The first two items in the list (output definition and data dictionary) are the result of a logical requirements definition procedure. The second two items (constraints and risks) are the result of a physical requirements definition procedure.

With this requirements definition procedure, the focus is still upon outputs (results), as it was in structured systems development. *The main objective of the requirements definition procedure is the clear, complete, and consistent definition of the principal outputs of the application or system.*

The requirements definition procedure is divided into two sequential phases: a logical definition phase and a physical definition phase.

Requirements Definition Procedure { Logical Definition Phase

Physical Definition Phase

In the logical definition phase, the systems definer is interested in the ideal (functional) system. This phase attempts to get at what any system of this type must do to support the basic application function.

Experience shows that, underlying any specific accounts receivable, inventory, or job-shop scheduling application, there is a functional flow of information, dictated by the type of application. Structured requirements definition, then, attempts to get at this underlying functional system through the logical definition phase.

But what about the myriad unique characteristics of each specific organization? Isn't it necessary to take these into account? Of course, but not as the first step. It is best to deal with the functional system first, and then, in the physical definition phase, to deal with those characteristics that are unique to the specific user.

By dividing our concerns, we also divide our problems. By first concentrating on the ideal functional system, we are developing an understanding of the part of the application that is most stable, and that must be present in any good system of this type. Then, once we have this stable functional definition, we can systematically consider those things that are truly unique about our situation.

LOGICAL DEFINITION PHASE

It is widely agreed that the best place to begin is at the beginning. Unfortunately, the right beginning is not always easy to find, and this is as true of systems development as of any other activity. Although the requirements definition process has been widely discussed in systems literature, most traditional approaches to defining requirements are intuitive, based mainly (if not solely) on the personal preferences of the author.

Most traditional systems approaches start with the recommendation to "study the existing system," which seems reasonable enough. Indeed, one would not want to undertake any new system without a clear understanding of the old. But what sounds reasonable in the abstract may not be so reasonable in practice.

The problem with studying the existing system is that you need to know what you are looking for before you can determine what to ask about the old system. And it takes considerable experience for most analysts to define successfully the requirements for a good information system.

For example, a junior analyst was asked to develop a new system. When the junior analyst asked how to go about it, a senior analyst suggested that he study the existing system and then automate it. In a few weeks, the young man returned in a very disturbed state. After watching the operation and talking with the personnel involved, he had concluded there was no hope of ever automating the existing system because, "There is no existing system. They never do anything the same way twice!"

This story illustrates an important point, one so

obvious that it is often overlooked by individuals with a great deal of experience: *the process of uncovering a system, especially in a traditional environment, is a skill that must be learned.* In this particular instance, it was probably not that there was no system -- the junior analyst simply did not know the right way to find it.

A number of serious questions are involved in successfully studying (or even documenting) the existing system in detail.

- How much detail is enough?
- How do we know when we have everything?
- How will we know if the current system is bad?
- How will we be able to separate information needs from wants?

In practice, it is sometimes easier to develop a new program from scratch than to analyze the old one, especially if the old one started out as a bad system. Indeed, it is not clear that a careful study of an existing bad system will ever tell you what a good one should look like. Thus, the blanket recommendation, "Study the existing system," can be the Bermuda Triangle of systems work, a place where many good systems are lost without a trace (not to mention a lot of good analysts and programmers).

What should we do then? Should we ignore the existing system entirely and develop an ivory tower idea of how a good system should look? Not at all. We just need to be careful, for, apart from the difficulties involved in studying an existing system, there are many sound reasons for systematically examining what currently exists.

For one thing, the question of politics comes into play. Computer analysts are often accused of being too concerned with the technical problems associated with an application while neglecting the organizational and human questions involved. If a systems definer spends a great deal of time interviewing users and getting to know the people and their jobs, his or her recommendations are apt to be considered in a better light and, clearly, establishing this credibility is an important factor in developing any successful system.

The structured requirements definition approach is to try to understand the system a little at a time so each piece of information gathered can be put into the proper place. Instead of doing a superficial survey in which the analyst attempts to cover all elements of the system once and for all by collecting as much information as possible for future analysis, in structured requirements definition, the systems definer works repeatedly with the users of the system.

The logical definition phase, then, defines the general **context**, works through to a careful definition of the various **functions** and tasks that must be accomplished, and delineates the information **results** required to support these functions and tasks.

Logical
Definition
Phase

{

Define the Application Context

Define the Application Functions

Define the Application Results
(principal outputs)

Defining the Application Context

One of the systems definer's principal concerns at the beginning of the definition process is to ensure that the context of the system is correct. On the one hand, an analyst wants to make certain the scope encompasses everything required; on the other, he does not want to include anything extra.

The bigger the project, the more important the process becomes. Too often, the definer opens the door too wide, too soon. Indeed, the easy way out is to let the project encompass what anyone might want--the so-called "kitchen sink" school of systems definition. Including as much as possible in the initial determination of the scope may ease the user's concerns in the early phases of the definition process, but this often turns a reasonable project into one of monstrous proportions. In many cases, the discovery of how big the project is occurs too late to react to it. So we set the context carefully, using the following procedure.

Define the
Application
Context

{

Define a user-level entity diagram (for each user)

Define a combined user-level entity diagram

Define the application-level entity diagram

Define the objectives

In developing the context, entity diagrams help us work with the users of the system to achieve an understanding of exactly what is included in the system and, conversely, what is not. We take individual views and combine them to produce an overall picture; then we establish the objectives.

Defining a User-level Entity Diagram. There are a number of ways to develop a good entity diagram, but here are the most helpful rules:

Develop
an Entity
Diagram

> Write the name of the organization, user, and application or system at the top of the diagram.
>
> Put the name of the user (participant) in a bubble in the middle of the diagram.
>
> Array the entities that the user will interface with in bubbles around the edges of the diagram.
>
> Draw and label arrows to represent interactions or transactions between entities.
>
> Review the diagram, and redraw if necessary.

We now have the framework for a procedure to define the application context, so let's follow an example to see exactly how this procedure works.

Suppose the management of Acme Arch Supports, Inc., has decided, on the basis of a systems planning study, to develop the requirements for an improved accounts receivable application. And suppose, further, that you have been hired as the project manager.

Structured planning, like structured requirements definition, is a procedure to help management identify essential, feasible applications, and to set priorities for them. The main product of the planning phase is a planning document.

In general, a systems planning document should identify: the real problems (opportunities), the users, and the scope that a specific application is to address. The chart below shows a summary of a planning document.

ACME ARCH SUPPORTS, INC.
PLANNING DOCUMENT (Summary)
Accounts Receivable Application Study

Problems/Opportunities

- Volume of business has doubled in last 4 years
- Billing is often delayed due to poor procedures
- Collection problems are increasing due to lack of accurate accounts receivable information and high interest rates
- Borrowing has increased due to expanded volume and poor billing and collection procedures

Users

- Accounts Receivable Section (Finance Dept.)
- Shipping Department
- Sales Department
- Credit Department

Scope

- Define a system that will have the capability to support increased volumes and have increased flexibility to deal with additional management reporting requirements--to replace the existing billing and collection system.

With the planning document as a basis, let us define the context. In this example, we will develop entity diagrams for the Accounts Receivable Section, the Shipping Department, the Sales Department, and the Credit Department. After we have developed these individual views, we will combine them into a user-level entity diagram, develop the application-level diagram, and define the objective.

In developing each of these diagrams, we will try to break things into simple steps, leaving as little as possible to chance. Therefore, we will go through the steps to develop an entity diagram for the Accounts Receivable Section in detail.

Step 1: ***Write the name of the company, user, and application or system at the top of the diagram.***

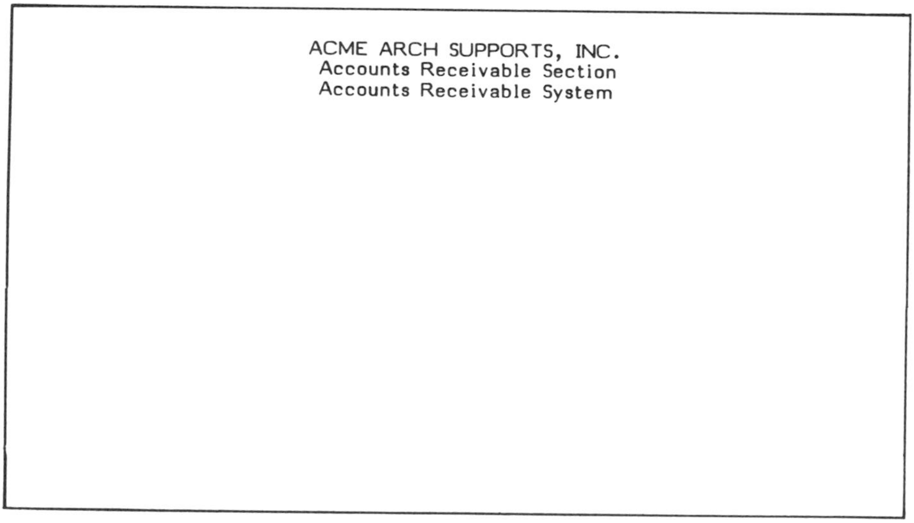

ACME ARCH SUPPORTS, INC.
Accounts Receivable Section
Accounts Receivable System

Figure 6.1.   Entity diagram - Step 1

Getting the right title improves the communication process. Limiting the universe narrows the possibilities to be worked on. Through the simple act of labeling the diagram correctly, we have identified the scope of the kinds of information we will be interested in, i.e., only those things relating to the Accounts Receivable Section of Acme Arch Supports, Inc., with respect to the accounts receivable function.

> **Step 2:**   *Put the name of the user (participant)*
>   *in a bubble in the middle of the*
>   *diagram.*

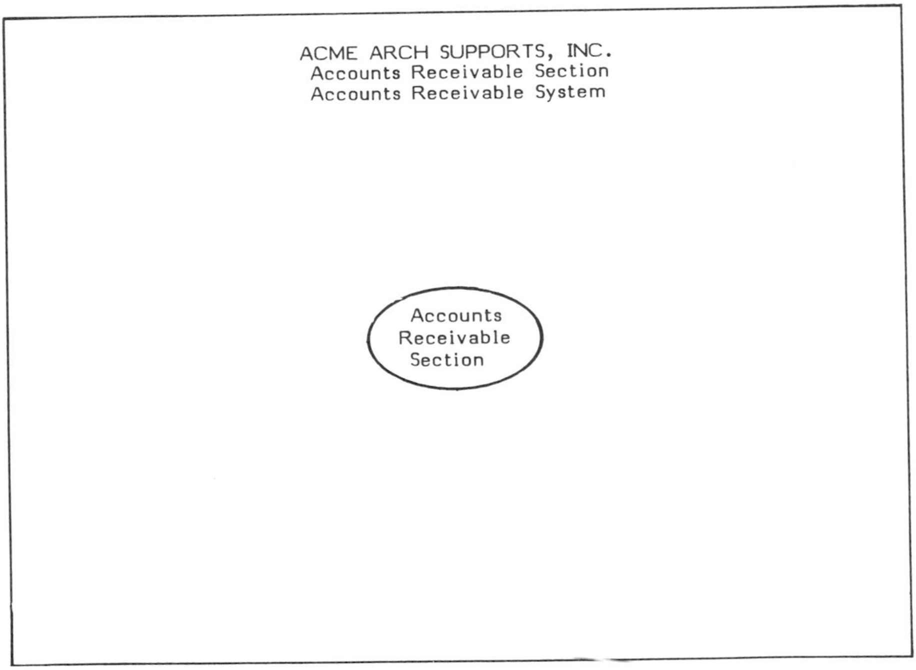

ACME ARCH SUPPORTS, INC.
Accounts Receivable Section
Accounts Receivable System

Accounts
Receivable
Section

Figure 6.2.   Entity Diagram - Step 2

Before the context of the systems activity can be correctly defined, it is essential to understand the system's audience. Since different organizations have different objectives, it is critical that the analyst clearly spell out the user for whom the entity diagram is being developed.

Step 3:   *Array the entities that the user will interface with in bubbles around the edges of the diagram.*

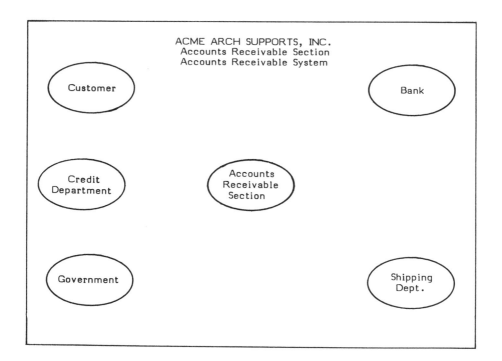

Figure 6.3.   Entity Diagram - Step 3

Developing a good entity diagram involves identifying the major internal and external entities that interact with the user in the context of this system or application. "External" is stressed here because systems are often developed as if they had no relation to the external universe. Resulting systems are technically correct, but ineffective, since they invariably leave out major considerations. Though they may be clear, they are incomplete from a business standpoint.

> **Step 4:** **Draw and label arrows to represent interactions or transactions between entities.**

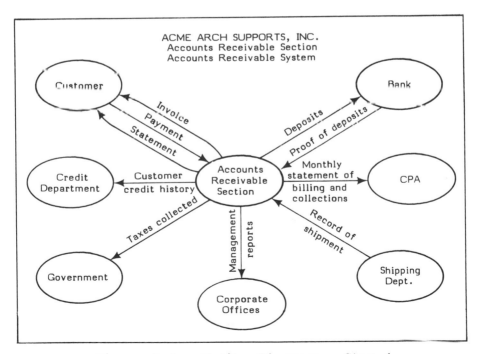

Figure 6.4. Entity Diagram - Step 4

Identifying the transactions is largely a free form activity. The basic transactions, such as invoices, payments, and customer statements, come easily. As the user examines the diagram, additional transactions get added. In fact, identifying new transactions often leads to identifying entities that have been omitted from the diagram itself. Here, for example, the Corporate Office and CPA entities have been added to the diagram produced in Step 3 because transactions were encountered to entities that were not included.

The user-level entity diagram is a means of gathering initial information through research and interviews. These diagrams provide a mechanism for getting started quickly and for capturing an overall view of the application quickly and in a fashion that leads to rapid user feedback.

Step 5:   *Review the diagram, and redraw if necessary.*

The first few versions of an entity diagram are apt to be messy, but that's all right. As we tell our students, it's not important if the first diagram is correct, but the last one should be. Therefore, after it has been agreed that all important interfaces for the user have been defined, some time should be spent on cleaning up the diagram.

Repeat the process (Steps 1-5) until an entity diagram has been drawn for each user in the system.

We now have a user-level entity diagram for one of the major users of the system, the Accounts Receivable Section. This represents a major view of the system, but not the only view. Junior analysts are often misled by their perception of "the user." It is a rare system that has only one user. In most cases, especially for large systems, there are a variety of users and user departments.

To guard against this problem, our procedure requires us to develop a separate entity diagram for each identified user department. Therefore, we would go through the same process for each of our other user departments. The final user-level entity diagrams are pictured in Figures 6.5 through 6.8.

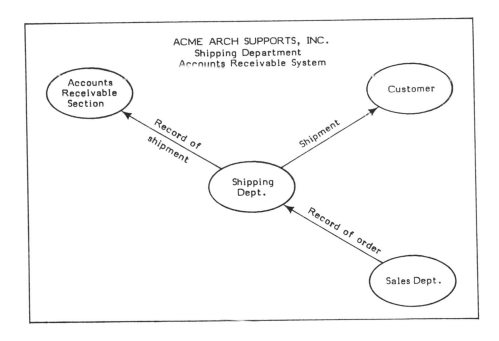

Figure 6.5. Entity Diagram (Shipping Department)

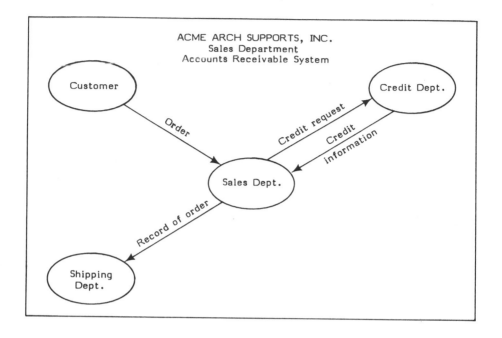

Figure 6.6.   Entity Diagram (Sales Department)

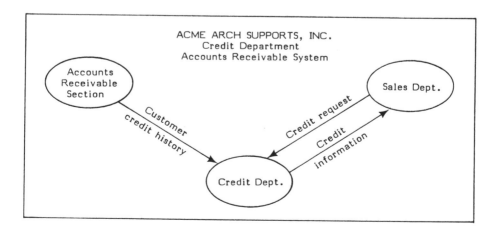

Figure 6.7.   Entity Diagram (Credit Department)

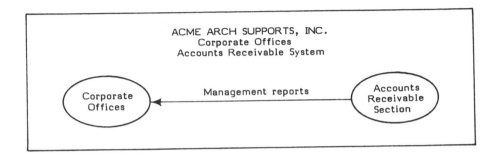

Figure 6.8 Entity Diagram (Corporate Office)

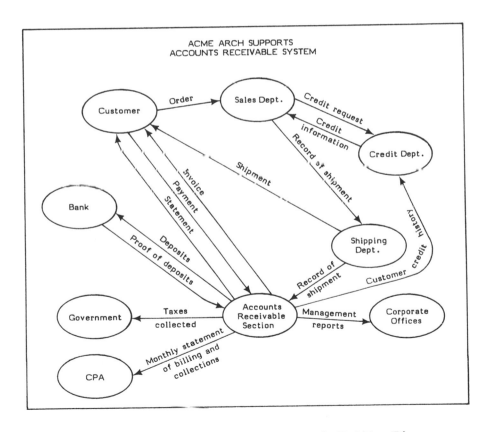

Figure 6.9.   Combined User-level Entity Diagram

Defining a Combined User-level Entity Diagram. We have individual pictures of what our application system has to do. Now we want to come up with an overall idea by combining all the diagrams into one (Fig. 6.9).

By combining the individual entity diagrams we may discover inconsistencies. For example, a transaction may be shown on one diagram that should also appear on another diagram. That is fine. Developing these diagrams can sometimes be an iterative process. Be sure to go back and correct any erroneous entity diagrams so all are compatible.

Defining the Application-level Entity Diagram. The combined user-level entity diagram contains a great deal of information--too much information. What we have in the combined diagram is a picture of all the interactions within the system. But what we would like to have is a picture of only the critical ones. This is achieved by defining the application-level entity diagram.

Defining the
Application-
level Entity
Diagram
{
Draw an application boundary around all internal entities to represent the application

Draw the application-level diagram, consolidating all internal entities into one bubble representing the company (organization, application)
}

From the external world, everything within our organization can be ignored. An organization may have

hundreds of branches, divisions, offices, etc.; however, from the outside it is just "IBM" or "General
Motors" or "the Department of Defense."

> *Draw an application boundary line*
> *around all the internal entities*
> *to represent the application.*

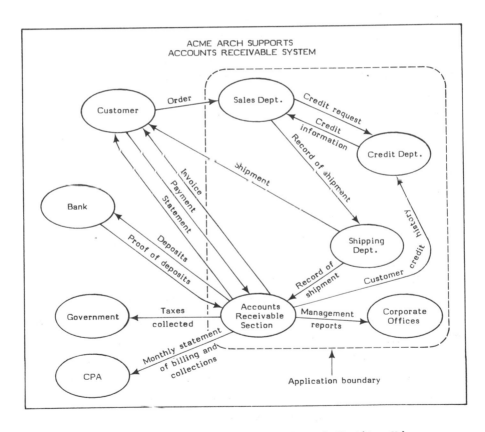

Figure 6.10. Combined User-level Entity Diagram
with Application Boundary

In the early days of structured design, there was a great deal of discussion about "top-down design." But one of the major problems was finding "the top." In order to establish "the top" for the purpose of requirements definition, we start with the pieces (individual user-level entity diagrams), add them together (combined user-level entity diagram), and then define the application boundary, i.e., the top (Fig. 6.10).

Drawing a boundary line around the internal entities identifies those entities that are within the application. This helps us identify the transactions that interface with the external entities which, in turn, helps us identify the major objectives of the system.

The application-level entity diagram (Fig. 6.11) "hides" the internal happenings of the system. This is intentional. To develop a truly "functional" system, anything that is not absolutely essential for the performance of the system must be eliminated from initial consideration. For this purpose, the internal flow of information can be ignored.

The application-level entity diagram provides the basis for developing consistent objectives and main line functions for the system. As a consequence, it is important that the diagram be complete and still exclude irrelevant information.

In the early applications of structured requirements, we attempted to develop the application-level entity diagram directly. We have found that, in practice, it is better to produce the individual user-level entity diagrams first, and then combine, determine the application boundary, and redraw to come up with the application-level entity diagram.

A good systems definer has the ability to grasp the essence of a system by sifting through great amounts of

> *Draw the application-level diagram, consolidating all internal entities into one bubble representing the company.*

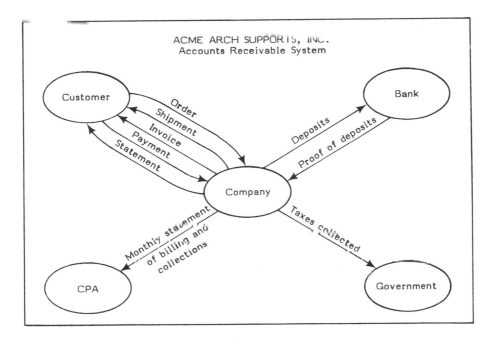

Figure 6.11. Application-level Entity Diagram

detailed information to come up with exactly what is required. To a degree, that skill is inherited, but part of it is learned, often the hard way. If we are to improve the quality of systems being built, we must be able to give inexperienced analysts and programmers better training, tools, and procedures. The entity diagram is one of the best tools yet devised for getting the right start in defining a system. It has proved to be a natural introduction to the rest of the requirements process.

In defining application context, we have combined, systematically, a number of interviewing, analysis, and documentation tasks normally done in a haphazard fashion in most traditional requirements definition processes. By developing a series of entity diagrams with the various users, we provide a natural vehicle for communicating the most important things a systems definer needs to learn at the beginning of a project, while avoiding the necessity of dealing with details that are not yet meaningful.

In order to understand and manage a larger application, it is vital to get "the big picture." But it is equally important that the big picture be complete and realistic. Our approach, then, is to protect ourselves from jumping to a premature "big picture" by first letting the user tell us what his idea of the system is.

In developing entity diagrams for various users, we combine these views and hide the internals. We end up with a bird's-eye view of the application that allows us to ignore, at least for the moment, how our organization currently solves the problem. That bird's-eye view is the application-level entity diagram. Together with the other entity diagrams, it represents a series of ways of looking at the application. Each view is significant, and, in its own way, a valid picture.

Next we need to define the application objectives. We need to determine the measurable things that the application must do to be successful.

Defining the Objectives. In structured requirements definition, we want to make everything in the methodology as explicit as possible. This is especially true in defining the application objectives. We want the objectives to be measurable and to follow directly from the analysis we have already done, so we use the application-level entity diagram as a basis for identifying the primary objectives.

We define the external objectives by inspecting the central entity in the application-level entity diagram --the Company in this case--and each point where a transaction (arrow) enters or leaves the Company. If an arrow is leaving the entity, then the objective can be stated as "send . . . transaction-x." If the arrow is entering the entity, then the objective can be stated as "receive . . . transaction-y." Out of this analysis, measurable, external objectives can be defined.

OBJECTIVES

1. Receive orders (from customers)
2. Send shipments (to customers)
3. Send invoices (to customers)
4. Receive payments (from customers)
5. Send customer statements (to customers)
6. Send deposits (to banks)
7. Receive proofs of deposit (from banks)
8. Send taxes collected (to governments)
9. Send monthly statement of billing
   and collections (to CPA)

By concentrating on the external relationships, the essence of a system becomes clearer.

The list of objectives is not very exciting. It doesn't talk about exotic goals and it doesn't deal with internal realities. But, as a first cut, it is useful to restrict objectives to direct, observable kinds of information. Whereas "develop integrated personnel data base" sounds good as an objective, experience indicates that such objectives are difficult to measure and, consequently, do not provide standards of success or failure for a project.

Much better is the objective that states "produce so many hourly payroll checks once each month." The more specific the goal or objective, the more productive the team working on the project proves to be. Highsounding goals are nice, but are often misleading. The more concrete we can be, the more likely we are to succeed.

So far in this example we've defined two important things: the system's context and the system's objectives. Now we are in a position to lay out the "main line" of the system.

Defining the Application Functions

At this point, the first major step of logical definition has been completed and we have a preliminary set of documentation that describes the context and the objectives of the application. The next major activity involves defining the application's functions. This procedure can be described according to the outline at the top of the next page.

Defining the "Main Line" Functional Flow. In defining the application context and objectives, we have fixed the boundaries of the application in terms of the

Define the
Application
Functions

{
Define the "main line" functional flow

Define the scope

Analyze the functional processes
(steps or subsystems)

Define the decision support functions
}

participants (entities) and the interactions (transactions). This tells us what the system has to do as a minimum.

Even though the entity diagram is a valuable tool for getting started, it is best used to present a static view of the application. To define the application functions, some way must be found to systematically turn that static view into a dynamic one, and for this we will use the "assembly-line" diagram.

We would like to have some sort of functional flow representing the highest level of the system. In some circles, this is called "the main line," the essence of a dynamic system. The main line shows the principal transactions and processes tied together in a stream.

In any system there is a central stream (or streams) of activity that is the key to the operation. All systems have such a primary flow, although it is often difficult to discern what it is. Developing the application-level entity diagram and listing the objectives simply provide the analyst with the means to discover the various interfaces in the main line without getting lost in details.

But how do you derive this main line? The classic method is to trace transactions through the system, based on user interviews. This approach, however,

often allows for extraneous steps in the system. The
approach we employ is to define the minimal activities
that have to happen, using the following procedure:

Develop
Main Line
Functional
Flow

Number the transactions, assuming
ideal conditions, beginning with the
transaction that starts the process and
following it through

Develop an assembly line for the primary
output(s) of the system by putting the
last numbered transaction(s) on the left
and working backward, transaction by
transaction, until each numbered
transaction is included.

As each transaction is included in the
functional flow assembly line diagram,
mark off the corresponding objective
from the list of objectives.

For any remaining transactions, start a
separate assembly line for each and work
backward until transactions occur that
have already been included in some other
assembly line process.

Add the separate assembly lines together
to create a total assembly line.

Let's see how this technique for finding the main
line works. First, we take the application-level entity diagram and number the transactions sequentially,
starting with the initial one. In this case, the initial transaction is clearly the order, so we number it
(1), and proceed to follow it through shipment (2),
invoice (3), payment (4), deposit (5), and proof of
deposit (6).

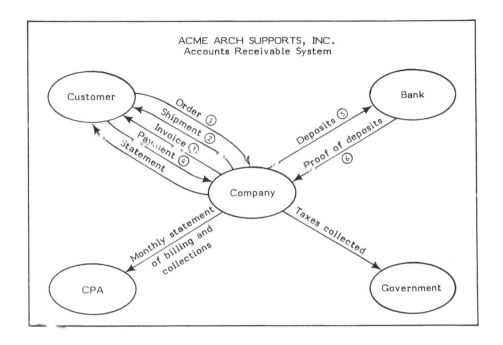

Figure 6.12

To develop the main line for this first stream, then, start with the last transaction (proofs of deposit) and work backward.

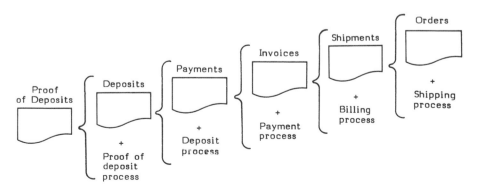

Figure 6.13. Main Line Functional Flow, Stream 1

From this view, it is clear that one transaction leads to another. All the initial and intermediate steps are done to achieve the objective: in this case, obtaining the proof of deposit.

It is not accidental that one step leads to another in a functional flow diagram. Although each of the steps (outputs) in the functional flow can be thought of as equally important, some (namely the ones on the left) are clearly more important than others. Most of the initial or intermediate transactions are a means to the ultimate product.

In the process of working backward to determine the main line, notice that we have effectively included most of our original objectives within the flow. Indeed, the main line covers the basic objectives, and these can be checked off the list of objectives.

OBJECTIVES

* 1. Receive orders (from customers)
* 2. Send shipments (to customers)
* 3. Send invoices (to customers)
* 4. Receive payments (from customers)
  5. Send customer statements (to customers)
* 6. Send deposits (to banks)
* 7. Receive proofs of deposit (from banks)
  8. Send taxes collected (to governments)
  9. Send monthly statement of billing
     and collections (to CPA)

We are not finished, however, until all objectives have been checked off. We must still produce customer statements, and send taxes collected and the monthly statement of billing and collections.

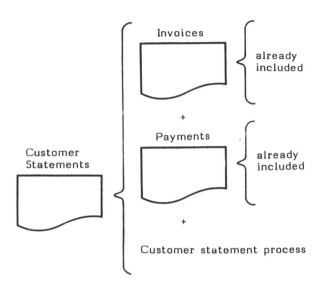

Figure 6.14.   Main Line Functional Flow, Stream 2

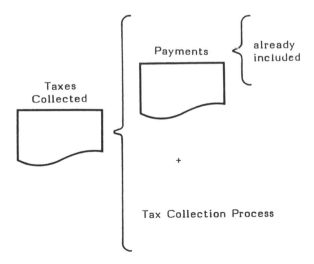

Figure 6.15.   Main Line Functional Flow, Stream 3

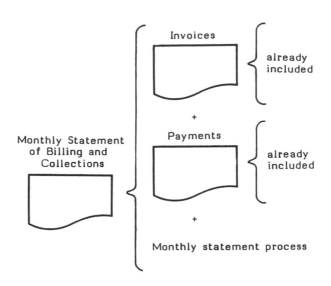

Figure 6.16.   Main Line Functional Flow, Stream 4

Now we have dealt with all the objectives.

OBJECTIVES

* 1.   Receive orders (from customers)
* 2.   Send shipments (to customers)
* 3.   Send invoices (to customers)
* 4.   Receive payments (from customers)
* 5.   Send customer statements (to customers)
* 6.   Send deposits (to banks)
* 7.   Receive proofs of deposit (from banks)
* 8.   Send taxes collected (to governments)
* 9.   Send monthly statement of billing
       and collections (to CPA)

Since we have handled all the objectives, we can build a total assembly line that represents a high-level, main line, functional-flow diagram (Fig. 6.17). Everything we must produce with this system is produced. Each step (brace) in the diagram represents a functional process that produces one primary (major) class of outputs only, although each step can, and usually will, produce any number of secondary outputs.

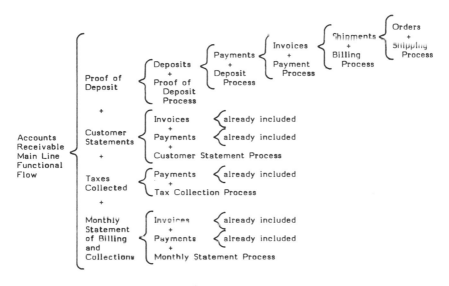

Figure 6.17

The main line ties everything together, showing the underlying causal relationships. We do **x** to get **y**, **y** to get **z**, etc. It turns the application into a system that makes sense. And while it represents an ideal situation (with all the exceptions left out), it also represents the minimal flow through the system.

Defining the main line is a critical element in developing a successful application. This is particularly true on very large systems, which may propose so many different functions that the main line tends to get lost in details. By consciously focusing on the systematic development of the main line, the requirements definition procedure materially aids the systems definer in understanding this principal thread.

Defining the Scope. In addition to tying everything together, the main line functional flow provides an excellent basis for defining the scope of the detail definition process. For many cases, the functional flow is still too large to be attacked, given our resources and time, and additional boundaries must be imposed. We call this **defining the application scope**.

Depending upon our problems, resources, and time frame, we may define our scope differently. For example, we may want to include everything on the main line functional flow as the scope of the application, or we may want to restrict ourselves to a subset of the functional processes, or even to a single functional process such as invoicing.

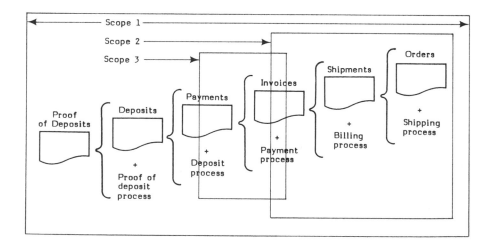

Figure 6.18

In general, it is desirable to set the smallest scope that will get the job done, but you have to be careful not to set the scope so small that you fail to solve the original problems defined in the planning phase. For our case study, we will take the entire main line functional flow as our scope, although we reserve the right to redefine the scope after we know more about the detail functional processes.

Analyzing the Functional Processes.   The main line functional flow is a general model of the system we're interested in. For most systems, it does not represent a detailed functional description of the application, but it does provide a good model on which to base further hierarchical decomposition of the application.

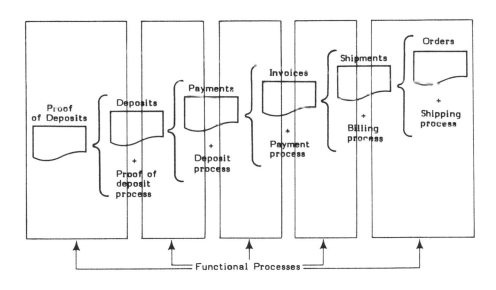

Figure 6.19.   Functional Processes

Each brace of the main line functional flow represents what we call a functional process. Since each transaction on such a diagram represents a class of outputs, each functional process is potentially a many-to-many mapping and, therefore, may need to be broken down into more detail.

A functional process may be a relatively simple, well-defined task, or set of tasks, in which case we refer to it as a **functional step**. In other cases, however, a functional process may be a complex set of activities, in which case we refer to it as a **functional subsystem**.

This is clearly an example of a recursive definition: a system is made up of functional processes that are either functional steps or functional subsystems, where a functional subsystem is simply a system at a lower level.

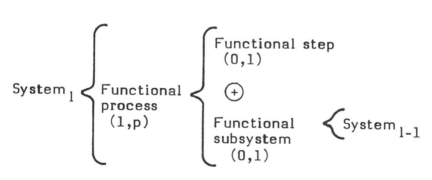

Figure 6.20

A large system, then, is first analyzed into its highest-level components (functional processes) and, if these pieces are too large to be dealt with as a single unit, the process of defining the functional flow is repeated, albeit with some variation. The pieces of the subsystem are, in turn, analyzed until they are small enough to be treated as a single unit, i.e., they represent functional steps. In other words, functional subsystems require the same kind of analysis devoted to the system as a whole--defining context, objectives, main line functional flow, and so on.

In our case study, we would start by studying the proof-of-deposit process, then the deposit process, then the payment process, etc. We suggest that you study the functional processes from left to right--from output to input. Clearly, given our output orientation, this should be no surprise. We find that the idea of working from outputs has a distinct advantage, even at the highest level of systems work.

Even where the application of output-oriented design techniques is accepted at the detail systems design level, systems analysts and designers have been slow to see the impact of applying that same approach at the requirements level. Large projects are usually broken down into phases and, invariably, the first phase to be defined is the input phase, e.g., order entry.

This input-oriented requirements definition causes a great many management problems. We start defining requirements based on inputs and, consequently, are left guessing about what the system should do. On the average, we guess wrong on a good many issues. As a result, when we move from one phase of a project to the next, say from order entry to shipping, we discover that we are faced not only with building a shipping process, but also with performing a major overhaul of the recently installed order-entry process.

Each process in a functional flow is constrained by the system it feeds. Therefore, going from left to right allows the project to move from goals to inputs, minimizing the amount of interaction in the project.

A word of warning is in order here. Most organizations are divided along functional boundaries. Even though the functions are part of a larger system, the managers and workers are often not interested in any-thing outside their own parochial concerns. Therefore, while studying the functional processes in a rational sequence from output to input is ultimately the most effective approach, a certain amount of education is required at all levels to convince management, users, and data processing personnel of the wisdom of this method. Sometimes, for the sake of survival, it is necessary to study and implement functional processes out of order, but management ought to be aware of the risks involved.

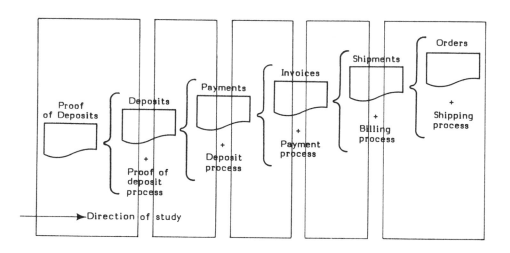

Figure 6.21.   Direction of Study of
Functional Processes Within a System

Analyzing Functional Subsystems.   We can define a functional subsystem just as we do a system, using bas-ically the same procedure and tools.   Figure 6.22 is the entity diagram for the proof-of-deposit process.

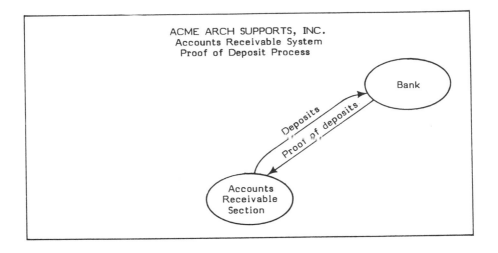

Figure 6.22. Entity Diagram

Using the same method of defining objectives discussed earlier, we can list the objectives of the proof-of-deposit process.

OBJECTIVES

1. Send deposits (to banks)
2. Receive proofs of deposit (from banks)

Next, number the transactions on the entity diagram in the order they occur (not shown) and develop an assembly line (main line) for the proof-of-deposit process.

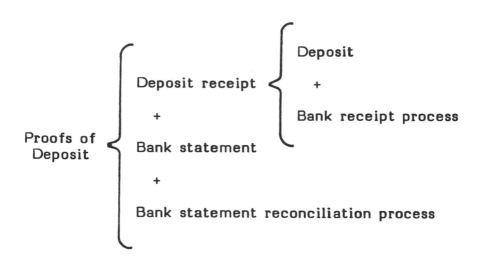

Figure 6.23

To this point we have employed the same approach on the subsystem level as we did at the systems level. There is, however, a difference. After we have developed the main line functional flow for the subsystem, we define exception or error handling. This is the "control feedback" of the system.

Defining the Control Feedback. The main line functional-flow diagram of a subsystem is predicated on the ideal circumstance--that everything will go right. But things don't usually work that way in the real world. Not only does everything not go exactly right, but according to Murphy's law, "Anything that can go wrong, will."

Therefore, any technique that sets out to develop systems based on the idea that everything will go right is doomed to failure. On the other hand, it is easy, when attempting to understand a system for discussion with users, to become permanently sidetracked with the exceptions and miss the main line entirely.

Structured requirements definition works at resolving this conflict one step at a time. First, we try to identify the essence of the system, the main line. Next we identify the processes within the main line and the ideal functional flow for each process. We then go through each process, systematically applying Murphy's law to identify the exceptions

In an assembly line diagram each brace can be thought of not only as a process, but also as a filter. Therefore, in the ideal functional-flow diagram only the good items pass through. In the shipping process, good orders become shipments; in the billing process, good shipments become invoices; in the payment process, good invoices become payments, and so on. When we apply Murphy's law, we ask, "What about the 'nonshipments,' the 'noninvoices,' and the 'nonpayments'--how will we know about those?"

By trial and error we discovered a way of systematically uncovering exceptions. First we ask the user to define the best (normal) case. For example, in analyzing the proof-of-deposit process, we discuss the case in which the deposit occurs without incident.

In Figure 6.24 we add a subscript to the deposit and deposit receipt to indicate the current business cycle (t). Therefore, we obtain proofs of deposit in cycle t for deposits that processed without any problems.

But there can also be proofs of deposit from previous cycles. Instead of dealing with all of these time periods at once, we take them in order, from the next-to-last business cycle (t-1) and work backward to successively earlier cycles t-2, t-3, and so on.

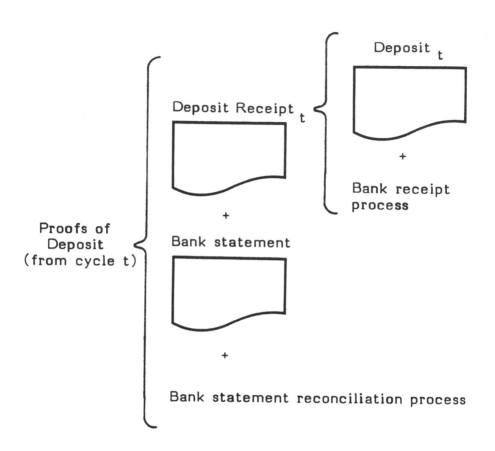

Figure 6.24. Functional Flow, Current Cycle

If you don't get it right the first time, you ought to get it right the next time around. In other words, a good system ought to be set up to be self-correcting. To get proofs of deposit in the current period that were in error in previous periods, we need a functional flow (Fig. 6.25) that accounts for those errors.

There are two things that can go wrong the first time around between you and the bank. You can make a mistake, in which case the deposit receipt should be revised; or the bank can make a mistake, in which case

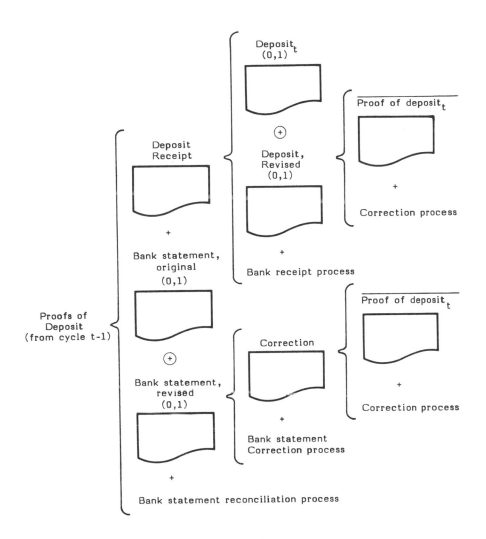

Figure 6.25.   Functional Flow, Second Cycle

the bank statement is wrong and should be corrected. The functional flow in Figure 6.25 handles those cases.

But we are not done.   What about those items that are still not reconciled after the second cycle?   Well, we try again.

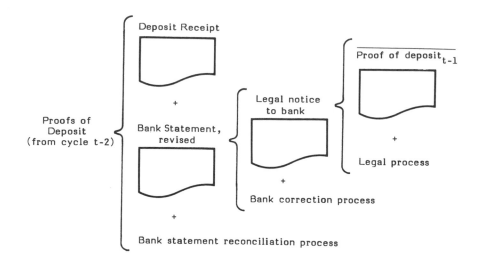

Figure 6.26.   Functional Flow, Third Cycle

In this case, we could not get the bank to see it our way, so we took legal action. As a result, we receive a revised bank statement that allows for a reconciled proof of deposit.

And we're still not done. What if we fail in our legal action, or decide not to press our case? We then obtain closure by admitting some (pseudo) proof of deposit for cycles greater than t-2, i.e., t-3, t-4, etc. (Fig. 6.27).

The systematic application of Murphy's law keeps repeating "but what if . . . ?"

> "What if A goes wrong?"
> "We do B."
> "What if B goes wrong?"
> "We do C."
> "What if C goes wrong?"
>
>   • • •

At some point, however, depending on the functional subsystems you are investigating, you have to put a stop to this process. In accounting systems, for example, there is always a mechanism for "writing off" a transaction that cannot be handled any other way. In structured requirements definition, there is always a last functional flow that covers all the loose ends.

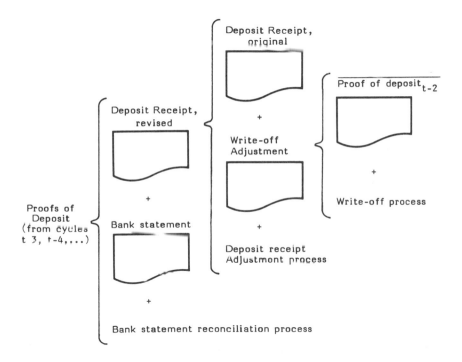

Figure 6.27. Functional Flow, Fourth Cycle

We have come a long way in our systematic exploration of the system. We've been able to uncover exceptions to our ideal (normal) process in a way that puts all exceptions into one overall subsystem.

Not only have we defined the "normal exceptions," i.e., those we must be able to deal with on a regular basis, we have also provided a means, via the "write off," for dealing with exceptions that occur only rarely. By doing this, we can guarantee that our subsystem has **closure**. Every deposit will result in a proof of deposit in one way or another.

Now that we have all the pieces, we can construct a functional flow for the proof of deposit functional process (subsystem) that contains everything defined so far (Fig. 6.28).

In general, management is concerned with negative outputs. Errors and omissions are bound to occur; therefore, the better the system, the more emphasis is placed on finding errors before they become large-scale problems.

The process of systematically applying Murphy's law is a more comprehensive way of dealing with exceptions than the traditional one of trusting solely the interviewing process. Murphy's law ensures that all the bases are covered. *It is not the things we get wrong in requirements definition that kill us; it is the things we don't get at all.*

Organizations try to hire analysts experienced in specific applications because they hope those analysts will be able to fill in the holes that users don't mention. Even if a user doesn't specify that he wants certain control reports, the experienced analyst knows which ones are needed. The process of feedback control analysis provides, for the first time, a systematic means of covering all the bases.

The concept of defining the ideal functional flow and then systematically exploring error conditions has proved to be exceedingly valuable in practice. The example we have used here is a simple one, but it illustrates how control feedback cycles can be used to handle expected errors.

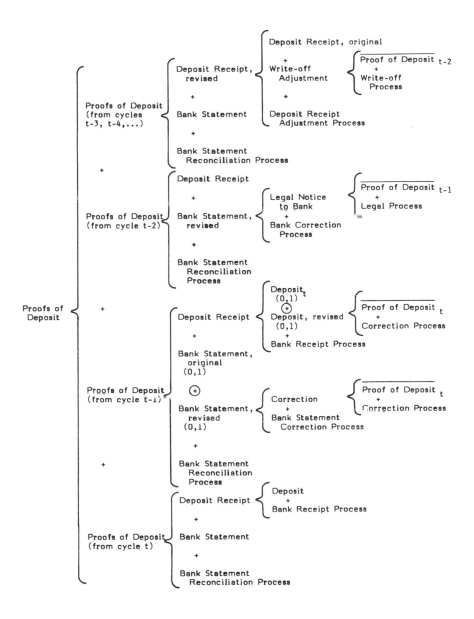

Figure 6.28

In a real application this process would be repeated for each functional subsystem. To conserve space we will only deal with the payment process. This is, however, a good example, since it is one where the feedback control is critical. People don't always pay on the first notice, or the second, or even the third.

The functional flow of the payment process covers all exceptional flows that result in the receipt of payments. It has, as you might expect, all of those means of correcting for nonpayment, but it also has an exceptional condition that was brought to our attention by one of our accounting friends: there is a positive (t+1) cycle for prepayments.

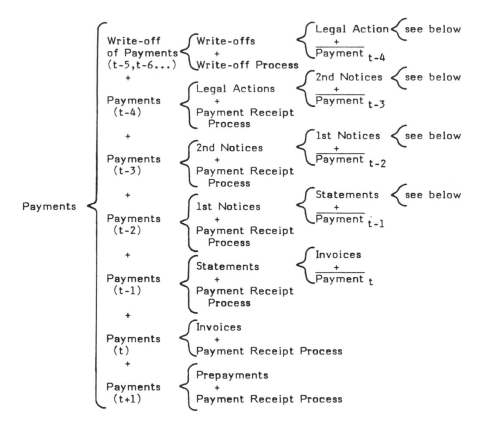

Figure 6.29

At some point, we are done with the analysis of the individual functional subsystems and, in all but the largest systems, we have only simple functional processes (functional steps) left to define. The definition of these individual tasks or procedures is an important part of the functional definition of the application.

Defining the Tasks and Procedures. The process of hierarchical decomposition may take place more than once, i.e., functional processes may themselves have processes. The basic rule is that the subdivision of systems into processes should be repeated until all the steps in the assembly line can be thought of as simple functional tasks or procedures, as activities that produce a single output or set of related outputs.

Structured descriptions of manual procedures are just as useful as structured programs. Many organizations use Warnier/Orr diagrams more for documenting manual procedures than for program design. (In the early days of structured design, we successfully used English in a structured form to document manual procedures, but we have found that the Warnier/Orr diagram is superior for the task.)

In defining the functional tasks or procedures, we are most interested in determining the following:

- What are the outputs of each task or procedure?

- What are the actions performed?

- What is the frequency of these actions?

- What information is required to perform the actions?

From this information, it is possible to develop a logical structure for each task. In the following example, we have defined a functional procedure, first in a narrative, and then in a Warnier/Orr diagram.

Late Payments Procedure (Narrative)

Each week the accounts receivable clerk will review all accounts for overdue payments and add any overdue invoices to the overdue invoice file. The overdue invoices should be sorted by amount (decreasing) within 120+, 90+, 60+, and 30+ days grouped within customer. The groups of customer invoices should be sorted by a total dollar volume. The accounts receivable clerk should contact each overdue customer. If the customer agrees to pay, a time for payment should be established and written on a follow-up report. If the customer does not agree to pay, then the reason should be entered on the follow-up report, and a time established for review. If the customer cannot be contacted, a note should be made on the follow-up report and the invoice replaced in the overdue invoice file.

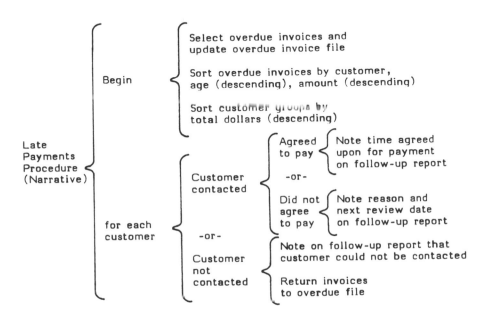

Figure 6.30. Late Payments Procedure
(Warnier/Orr Diagram)

Defining the Decision Support Function. Management functions are often characterized as: planning, executing (operating), and controlling. To this point the focus of structured requirements definition has been primarily on execution, and on the portion of feedback and control that is basic to the operational process. This initial concentration on operational feedback and control defines systems that are, as far as possible, self-correcting.

The best systems are built on a solid operational foundation. This satisfies the operational users of the applications, i.e., those persons charged with day-to-day activities--entering orders, shipping goods, issuing bills, collecting payments, and so on. Once the operational functions are defined, we add the truly managerial ones: planning and control. These are often referred to in systems literature as the decision support functions (DSF).

The decision support functions of any system or application produce two classes of results: planning results (outputs) and control results (outputs). And each class of results can be broken down into two sub-classes: predefined and ad hoc.

Define the
Decision
Support
Function
{
Define the management control function (predefined results)

Define the management planning function (predefined results)

Define the ad hoc reporting function
}

Defining the Management Control Function. Each organization is different, and each management has a different style. Within the ranks of management, a great controversy has raged over which is the most effective method: exception reporting or detail reporting. Often the choice has to do with the level of management and the relevance of the details. The higher one goes in the organization, the more emphasis is placed on management by exception. The closer one is to the actual operations, the more important the details.

Within the decision support functions, structured requirements focuses on those control functions that management users of the system deem most important. For example, in a tightly controlled governmental organization, the emphasis may be on monitoring costs vs. budgets; in a fast-growing organization, it may be monitoring cash requirements vs. cash availability; and in a development organization, it may be on measuring project plans vs. actual performance.

In general, the focus is on data and results related to a few key indicators. Indeed, most studies of information usage indicate that successful managers select a few key variables, and then monitor them closely. When any of the key indicators vary beyond certain limits, the good manager steps in to determine why.

Identification of key variables, then, is the task of management, particularly top management. Using these key variables as a base, the systems definer can work with the manager to formulate the predefined results (outputs) that will monitor and control the variables.

Defining the Management Planning Functions. Managing a large organization is akin to being the captain of a large ocean-going vessel. Because of its size, such a vessel cannot change direction rapidly. Therefore, it is important to map the course carefully. For the most part, experienced captains prefer taking well-traveled routes in favorable weather.

But conditions are always changing. So managers, like ship captains, must constantly review where they are and where they're going. Essentially, planning involves setting models and asking "what if" questions. "What if we sailed at different times?" "What if we carried different cargo?" "What if we used different fuels?" "What if we used different vessels?"

Whereas control reports normally interface directly with operational data, planning results often require additional data. Planning models may use operational data to determine parameters and trends, but planning functions, by their very nature, go beyond the current environment, pulling together data from other sources. For example, an increasing number of commercial data bases are available to support planning. Econometric data bases provide forecasts of various economic trends by region and time; technological data bases forecast technological trends by field and time, etc.

In the rest of the system, i.e., in the operations and control functions, the functions remain relatively constant and the data values vary. In planning, the opposite is true. Planners will build a planning data base, then process it many different ways, depending on variances in the key parameters.

If the organization is large and the planning model is complex, requirements definition for this function must be extensive and, as much as possible, explicit. Planning is similar to research because it is a speculative process, and planners are notoriously reluctant to commit themselves to specific results. Good planning requires a specification of the goals. Whenever usable, stable outputs can be identified, the entire systems process is facilitated.

The systems definer should be prepared to be tolerant when dealing with the planning functions, for what might appear to be a ridiculous requirement for an operational or control function may be perfectly reasonable in the planning context. Further, the systems definer should be careful not to exclude certain planning functions because of difficulty in obtaining the inputs. Inputs to planning functions need not come solely from the operational system; planning data may be obtained through statistical sampling and merged with operational data to produce reports accurate enough for planning purposes.

Defining Ad Hoc Reporting Functions. A great many planning and control functions cannot be predefined. Situations arise that cannot be anticipated. Therefore, any advanced application must be capable of producing ad hoc reports when necessary. In this context, ad hoc is defined as permutations and combinations of data already existing in the system.

Most current computer systems provide powerful report generation and query language facilities to support ad hoc reporting. If such facilities do not exist for the applications being defined, they may have to be designed and developed as part of the implementation of the system.

The key to ad hoc reporting is the logical design of the systems data base. We have learned how to design and structure logical data bases to facilitate flexible reporting, providing the capability to produce reports. Structured systems development now includes a logical data base design process that produces what we call "logical bases files." The logical bases files mirror the requirements view by dividing the data base into entities, transactions, and cycles/events. Logical keys are placed on files so the transaction bases files can be selected, sorted, and merged to provide different views of the data.

In traditional systems building it was common to ignore the necessity of providing for ad hoc reporting. As a result, managers had to wait for long periods while special-purpose reports were produced, often at a high cost. In recent years, as general-purpose data bases and report writers were developed, the pendulum has swung the other way. Often managers are told not to worry about defining their results because the system can produce anything anytime.

Although this latter view has corrected many shortcomings of traditional systems, it leads to future problems because it is impossible to structure a sound system without first defining the principal results (or outputs). Once the principal results are defined, it is possible with today's technology to produce a wide variety of ad hoc results with little system effort.

Much of the current literature on systems development revolves around management information systems. But in most large applications, managers represent only a small portion of the total user population. And, in general, good management information systems rest upon good operational information systems. The stable system of the future must be balanced between operational efficiency and managerial flexibility.

In structured requirements definition we are most interested in providing a systematic means of defining applications to meet all the needs of the organization. By addressing management's decision support functions after defining the basic application functions, structured requirements definition ensures that management reporting is explicit, done at the right time, and based on a thorough understanding of the logical processes being managed.

The final functional specifications of an application include a set of assembly line diagrams of each functional system or subsystem, plus a Warnier/Orr diagram of each detail functional step (task or procedure). At this point, we have a set of documents that allows us to discuss the context and functions of the application. What is left is the precise definition of the results of the application.

Defining the Application Results (Principal Outputs)

Of all the parts of logical requirements definition, the definition of the results is the most critical. In defining these results, we must:

Define the
Application
Results
{
Identify the principal outputs

Define the principal outputs

Define the organizational cycles

Identifying the Principal Outputs. Before you can define something, you need to identify it, so the first step in the definition of the application results involves identifying the principal outputs of the application. It is critical to structured requirements definition to understand clearly what is meant by a principal output. *A principal (or primary) output is the reason for which we are running the system.*

There is often confusion among experienced analysts and users about this distinction. Examples of principal outputs are management reports, checks, inquiry response screens, and tapes, disks, or cards required by other external systems.

Things that do not represent primary outputs are
called **secondary outputs**. Examples of secondary outputs are control reports, data base dumps, edit list-
ings, and input screens. Secondary outputs are not the
reason we do the system. And in structured requirements
definition, we want to be sure that we are dealing only
with the minimum that the system must do.

To identify the principal outputs, we produce an **in-
out diagram** for each functional subsystem and, if need
be, for each functional step as well. The in-out diagram provides a simple view of a process in terms of
its external interfaces. It gives us an exhaustive
list of outputs that have to be defined in order for
the application to be completely defined.

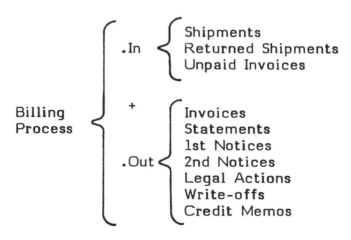

Figure 6.31

Defining the Principal Outputs. Outputs are intended
to support tasks, procedures, and decisions. Like
everything else in structured requirements, the definition of the outputs is (or ought to be) a natural out-
come of the functional flow, control feedback, and

functional tasks. If you know what your tasks and decisions are, then it is a relatively straightforward
job to define what the outputs should contain and how
often they should be produced. Even with careful work
to this point, however, we still discover a great deal
when we sit down with users to define outputs.

In a sense, when we start defining outputs, we are,
for the first time, creating a visible model for the
user. Since the outputs represent a large part of what
the user will employ to determine the practicality of
the system, it is critical to define them in the right
way. Every output in the system should be defined in
four ways:

Define
Output
{
Define output form
(logical data layout)

Define output content
(mook-up, sample, or simulation)

Define output structure
(Warnier/Orr diagram)

Define data elements
}

The strategy for definition varies, depending on
whether the output involved already exists. If it does
exist, a real example can be used for the sample and
the logical data layout can be produced by simply placing a sheet of tracing paper over the sample and drawing buckets wherever variables or literals occur.

Existing outputs should not be taken as given requirements. It is important for existing outputs to be

compared with the description of the tasks they are to
support. If they don't do the job, new outputs should
be defined. By carrying forward old, outmoded outputs
from one system generation to the next, we carry forward old problems as well.

If the output we are defining does not currently
exist, then the best technique is to develop a logical
data layout first, using the task or procedure definition as a guide. For example, if we are defining the
output to support the "late payment procedure," it
might look like Figure 6.32.

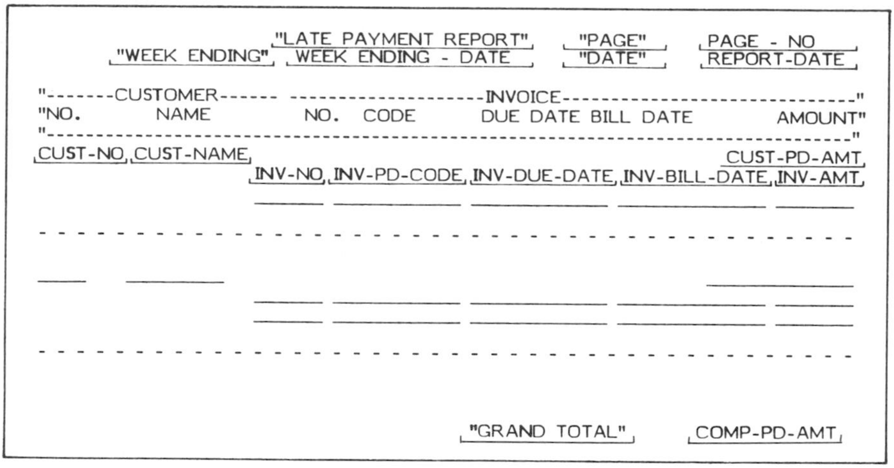

Figure 6.32

Based on the logical data layout, the Warnier/Orr
diagram for the structure of the data on the output can
then be developed. The best approach for coming up
with a good Warnier/Orr diagram is to write down the

data elements on an output definition form, along with the frequency of occurrence (1/universal) and the computation rules (Fig. 6.33). A universal is the name given to a set of actions or data. (At this stage it is recommended that you ignore headings and footings in the definition.) From this information you can define the frequency for each universal.

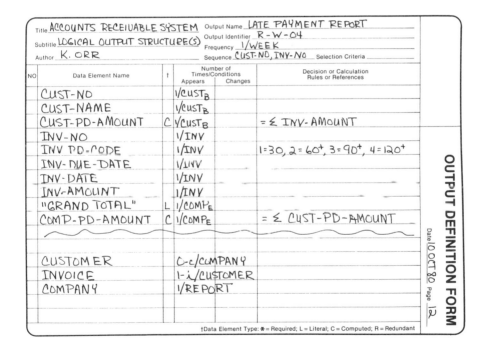

Figure 6.33

The Warnier/Orr diagram on the next page is easy to construct once you have completed the output definition form.

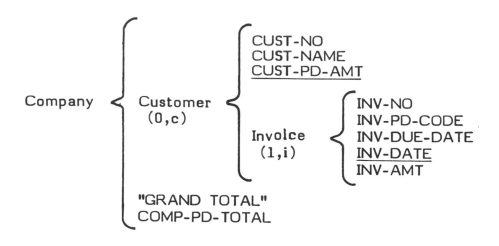

Figure 6.34

With the logical data layout and the Warnier/Orr structure diagram in hand, the construction of the sample output is straightforward (Fig. 6.35).

```
                    LATE PAYMENT REPORT              PAGE - 01
                    WEEK ENDING - 12/07/79           DATE - 12/08/79

--------CUSTOMER------  -------------------INVOICE----------------------------
NO.      NAME          NO.   CODE     DUE DATE BILL DATE        AMOUNT
--------------------------------------------------------------------------
0425   SMITH INC.                                               278,493.75
                       78321   4      79-01-15  78-11-13         20,500.00
                       79044   4      79-05-01  79-03-21         11,150.00

- - - - - - - - - - - - - - - - - - - - - - - - - - - - - - - - - - - - - -

0863   RANDALF BROS                                            223,621.80
                       78032   4      78-04-22  78-03-10         10,175.17
                       78061   4      78-05-22  78-05-03          3,216.25
                       79535   2      79-09-25  79-08-03          8,975.00

- - - - - - - - - - - - - - - - - - - - - - - - - - - - - - - - - - - - - -

                                      GRAND TOTAL              2,475,893.37
```

Figure 6.35

This seems like a lot of work to define an output. And if you have several hundred outputs, just defining them can be a monumental task. Tediousness aside, it still needs to be done.

On one project we were involved in, the requirements study had been done but the outputs had been specified only in terms of the traditional X's and 9's. Each output, consequently, was defined again in terms of Warnier/Orr diagrams and samples. There was a complete change in the way both data processing and the user viewed the results. Even after several hundred outputs had been defined, the project team still learned as much on the last output as on the first.

It is not unusual to go through many versions of outputs before users can agree on what they want, which can be frustrating. On the other hand, I still remember one user remarking, after we had reworked a set of outputs for about the sixth time, "You know, I'm sure glad we didn't implement the outputs we initially asked for. If we had, we would have been back in six months to redo the entire system."

Gerry Weinberg has said, "Paper changes are cheap." Nowhere is this more true than in the definition of the outputs of a system. If the outputs are wrong, everything else is apt to be fruitless. Extra time spent in requirements definition saves time in design, installation, operations, and, most importantly, in use.

Defining the outputs provides us with an excellent opportunity to develop a data dictionary at the same time. Indeed, if we don't do so, we'll be sorry when we discover we don't understand what was required on the outputs. (See Chapter 4 for a discussion of data dictionaries.)

A word is in order regarding the output we developed. Notice that the total amount past due for the customer (CUST-PD-AMT) is printed at the beginning of each set of customer invoices rather than at the end. This is not the normal way a data processing analyst would design an output, especially working in a traditional manner. But since we are working with a structured description of the functional task that needs to be performed, and we have specified that the customers appear in sequence beginning with the one owing the largest amount, then the report sequence is natural.

A final note about defining data elements: **Not all data elements are equal.** Indeed, in every application, there are a handful of critical data elements. *Often the definition of the calculation and decision rules for a single data element may represent as much as 40 to 60 percent of the entire application.*

In a real accounts receivable system in which we were involved, we discovered, much to our dismay, that a single element on the invoice ("price/unit") was the key to the entire billing system. This discovery was made very late in the requirements process. As a result, we are now careful to ask others who have worked on similar systems exactly what the critical variables are and then allocate our time appropriately.

Defining the Organizational Cycles. We have finished defining the principal outputs of the system. The only thing left to do in logical requirements definition is to fit these outputs into a total picture. Let's return to the functional-flow diagram and add two arrows, signifying feedback to control nonpayment.

The feedback loop (Fig. 6.36) associated with the customer statements, customer late notices, and the late payment report is intended to improve the payment process by reminding the customer of his current obligation. The customer aging report serves to alert management to serious problems and, along with the late payment report, provides feedback to the order acceptance process. If customers stop paying, we had better stop taking their orders.

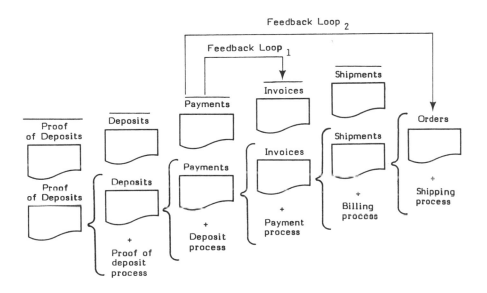

Figure 6.36

Systems are, by their very nature, cyclical; so the last part of requirements involves putting the individual system's output and input transactions into a total cycle diagram. Once again, the guide to that process is the frequency with which these outputs are required.

OUTPUT	FREQUENCY
ORDERS	50/day
ORDER STATUS INQUIRIES	750/day
ACKNOWLEDGMENTS	48/day
SHIPMENTS	240/week
INVOICES	235/week
PAYMENTS	42/day
CUSTOMER STATEMENTS	1200/month
CUSTOMER AGING REPORT	1/month
LATE PAYMENT REPORT	1/week
CUSTOMER LATE NOTICES	450/month
DEPOSITS	0-1/day
RECEIPTS	0-1/day
REPORT OF TAXES COLLECTED	1/quarter

From this information, we can construct a systems War-nier/Orr diagram (Fig. 6.37).

To this point we haven't addressed many of the technical questions data processing professionals seem to want to talk about, questions about on-line vs. batch, distributed vs. centralized, data base vs. nondata base. We have discovered that if we hold off talking about these questions until the physical requirements definition phase, and just find out what is needed, the decisions are usually made for us. There are still plenty of technical trade-offs to occupy the gurus, but the longer you can keep your attention on the logical problem, the easier it is to come up with the right solution.

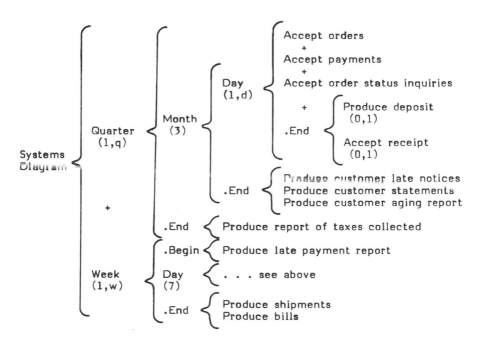

Figure 6.37. Warnier/Orr Systems Diagram

PHYSICAL REQUIREMENTS DEFINITION

When the logical requirements definition phase is completed, the systems definer has an exhaustive documentation of the application context, functions, and results. This documentation represents the essential things that the application must do.

In general, the logical requirements are fundamental underlying needs. Any good system must be able to meet the logical requirements, which also represent the ideal case. But where the logical requirements represent the ideal case, the physical requirements represent constraints upon that ideal case.

In structured requirements definition, the physical definition phase is divided into five steps:

Physical Definition Phase

- Define the constraints
- Define alternative physical solutions
- Define benefits/risks for each alternative solution
- Select the recommended course of action
- Prepare final requirements definition document

Defining the Constraints

A constraint is a limitation, and in requirements definition the critical constraints have to do with volumes, response times, frequencies, sensitivity and security of data, reliability, hardware, software, personnel, organizational considerations, time, and, finally, money.

Every system has its constraints, some more severe than others. For example, volume is a critical parameter. A system that makes sense for a local hardware store will clearly fail for a major manufacturer, even though the basic applications are the same.

Too often, generalized software is developed and promoted for use in a variety of organizations that would be more applicable to an organization of a particular size. We have discovered in recent years a vast difference between toy systems and what we have come to call industrial grade systems.

Toy systems are often deceptive. They take the required inputs and produce the desired outputs, but they work best for very small files. **Industrial grade systems** not only produce the right outputs from the right inputs, they do so reliably. Additionally, they are designed in such a manner that they can deal with large volumes and be used and maintained for a long time.

Volume and response time constraints often affect the frequencies by which various outputs and inputs are processed. If the volumes are low and the response times are long, the frequencies can be somewhat infrequent (weeks or months); but if the volumes are high and the response time is immediate, then frequencies have to correspond.

Reliability and security may determine the nature of the application, as well as what portions of the application are entrusted to the computer at all. If we are developing a batch system, we may be able to ignore certain considerations of reliability and security. On the other hand, if the application is a critical business function, and we are planning to put that function on-line, reliability and security have to be addressed.

Hardware and software considerations may influence how extensively we treat our application. Certain things may be easy to do if we have good data base management software, and difficult to do otherwise. If our hardware is restrictive, we may be locked into it for the initial version of this application.

Personnel and organizations may affect the final appearance of the application. Management, for example, may have definite ideas about the mode that best suits the organization. The type of personnel involved may require that we tailor the application to suit them. In recent years, much discussion has centered around the idea of "human factors engineering." The idea is to fit the application to human beings in general and to our personnel in particular.

Too often, systems are built without the proper consideration given to how they will be used in practice. Many systems are built around available technology, but are inappropriate to the class of user if, in fact, the class of user is truly defined. A great deal of thought must be given to the kinds of uses the application will have and the people who will be involved.

Time and money are also constraints, perhaps the ones that jump to mind first. Clearly, we want to give the user an application that works within the boundaries of the time and resources available. Often, after the logical requirements definition process has been completed, it is necessary to set priorities based on how much time and how much money we have.

In physical requirements definition, the systems definer attempts to spell out constraints that exist for the application, and to determine what impact those constraints are apt to have on the finished product.

Defining Alternative Physical Solutions

There is more than one solution to most problems. Today, with the wide variety of hardware, software, and application packages available, one of the most difficult processes is limiting the number of alternatives to be considered.

Once the constraints are defined, we recommend that the systems definer work with the user and appropriate vendors to define a set of alternative solutions that meet the basic logical requirements and constraints.

One should always consider the alternatives **do nothing** and **buy**.

There is an old adage that says, "If you can't make it better, don't mess with it." That is still good advice. In some cases there is no great economic advantage to computerizing an application. Where that is true, the systems definer should be wise enough to recommend that the systems development process be abandoned so critical resources can be applied to something more important.

An alternative that must increasingly be explored is the possibility of purchasing the application from some other organization. Each organization, as we have said, is different. Those differences, however, may not be as great as you would think.

As software development becomes more expensive and more critical, more organizations are buying packages from others or at least exploring the possibility. If an organization specializes in a particular type of application or in a particular industry, they may have developed a package that could meet your current needs, and have the capabilities to change for growth.

Purchasing packages is not without its risks. Sometimes packages are developed for a specific user, then generalized for sale, with an end result that is unwieldy or inflexible.

The most severe problem with acquiring packages is that they are often purchased by organizations that have not done a requirements definition of the application they think they are buying. Without a clear idea of what you need, you are at the mercy of the salesman.

In case after case, where organizations have gotten into trouble in purchasing systems from outside vendors, the problem started early. One justification for purchasing software is often to avoid going through the requirements definition process. Unfortunately, the result is that the organization ends up choosing between the packages that are offered, even if none of the available ones meet the real requirements.

Without a good requirements definition, it is impossible to determine how much modification will be required to adapt the application to the organization. Even the best package requires adaptation and, on very large systems, the cost of that process may run into six- or seven-digit figures.

Along with the alternatives of doing nothing or buying the application, we should come up with solutions based on building our own. In general, this should not be a single alternative. If possible, management should be provided with multiple alternatives to explore.

Defining Benefits and Risks

Each day in the media we are made aware of major proposals and major disasters. We hear about nuclear power plants that fail and will take billions to decontaminate and repair. We hear about space shuttles that are years behind schedule and billions over budget. And each day we hear about proposals to spend billions more on difficult decisions.

What is amazing is not that failures occur, but that we learn so little from them. We keep building more complex singular things, only to be surprised that we are crippled when they fail.

Most managers are familiar with cost-benefit analysis. They often fail, however, to deal with risk analysis. If the project fails, what will that do to us?

What if the project does not come in on time? Is it worth taking a chance on developing a new system predicated upon an untried technological breakthrough, especially if the chances of failure are high? Can we pull off this application with our personnel?

A careful risk analysis provides weights that can be applied to alternative solutions to give management more information on which to base judgments. In the past, data processing had to sell its ideas in order to get a chance. But today users and managers don't need to be sold; they want to apply the computer to every problem--yesterday. What is needed in this environment is a dose of common sense.

Risks may be outweighed by benefits. Managers are used to taking risks; indeed, that is what they get paid for. The systems definer must make clear exactly what those risks are. If management chooses a riskier project over a safer one, then that is its judgment, but it is the obligation of the systems definer to point out the risks.

Selecting the Recommended Course of Action

When all the information has been processed, some decision has to be made. Normally, one solution must be picked over the others. In most cases, this is a management decision; however, the systems definer or project team may have a great deal to say about which alternative is picked.

A selection process, unless all the weights favor the same alternative, requires consideration of trade-offs. Cheap solutions are often risky or low in overall benefits. Safe solutions often take longer and cost more than management would like. Industrial grade systems are more expensive than toy ones.

The process of selecting a course of action is a normal business function--it is carried on every day. Structured requirements definition simply provides the decision maker with the appropriate information to make rational decisions.

Preparing the Requirements Definition Document

The last task in structured requirements definition involves producing a final document describing to management, users, and the design team exactly what was discovered in the requirements process. This involves summarizing, cleaning up, and organizing the material that was developed in the course of the logical and physical requirements process. Below is an example of the outline of a complete structured requirements definition document (Fig. 6.38).

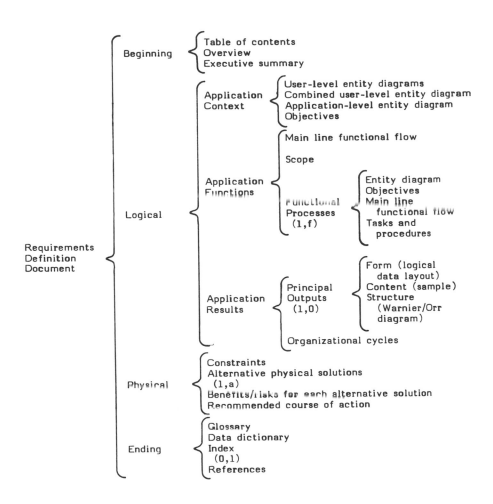

Figure 6.38
Requirements Definition Document Table of Contents

Management/7

****

Great advances do not come out of systems
designed to produce great advances.

Furthermore,

Complicated solutions produce complicated
responses (not solutions) to problems.

John Gall
Systemantics

As the development of good information systems becomes more critical, it is imperative to establish systematic procedures for all phases of systems building.
On the surface, at least, it would appear that systems
scientists ought to be good at this since they spend a
great deal of time defining precise procedures for machines and for other users.   But the evidence shows
that it is difficult to structure your own job, perhaps
because you are too close to what goes on, and because
you generally have little time for self-examination.

Procedures are always related to the available technology.   They change over time as new theoretical discoveries are made, as new problems or opportunities
arise, and as new tools become available.

For instance, only a few years ago open-heart surgery was rarely performed, usually without much hope
for success.   The surgery was delicate and the time to
perform the operation was exceedingly short.   As a result, the surgeon had to be particularly dexterous and
use every shortcut.

The invention of the heart-lung machine, however,
allowed the heart to be taken off-line for an extended
period.   Since the surgeon had much longer to perform
the operation, the open-heart procedure became more
detailed, and the results more favorable.   The additional time made it possible to introduce more steps;
the additional steps contributed to the increased success of the procedure.

Clearly, procedures and tools go hand in hand.   But
what about the relationship between procedures and man-

agement? Any critical procedure needs some external quality control process, some management requiring that practice coincide with standards. These management processes include spot checks, certification, retraining, and peer review in addition to the usual management controls of resources such as people, equipment, and monies.

As procedures change, training and management must also change to reflect the new activity. One of the frustrations in systems development is that management is often out of phase with technology. Many current project-management guides were produced by people who have not written a program or developed a system in ten or fifteen years, and if they haven't kept up with the technology, their approach may actually be counterproductive. Approaches that were effective for batch systems on second generation equipment may not be appropriate for on-line, distributed, data-based systems.

A good systems development methodology, then, is a precise, delicately balanced combination of theory, tools, procedures, training, and management. The creation of such a methodology is difficult because so many issues must be brought together.

The structured systems development methodology differs from others most strikingly in its point of origin (see Fig. 7.1). Many standard methodologies were initially developed from the standpoint of project management rather than from that of design philosophy. This is understandable, since any organization faced with creating large systems has to develop project control techniques to monitor the process. These methodologies, however, are often deficient in detailed technical directions. They represent a manager's idea of what a programmer or analyst ought to do.

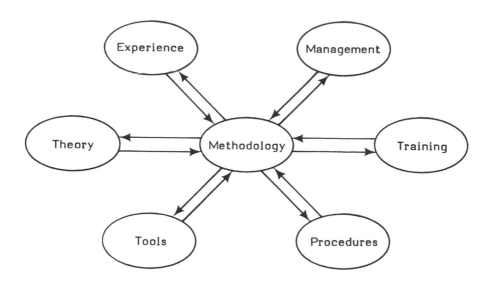

Figure 7.1

I became interested in project management in the late 1960s. Indeed, I had tried all the prescribed traditional approaches only to have them run aground upon the rocks of design and implementation. Thus I learned that project management can only propose; it lacks the power to dispose. In the face of inexperience or incompetence, it is powerless. When the "wrong" procedures are used, advanced project management techniques may even be dangerous.

Moreover, if project management is the starting point in the development of a methodology, the tendency is to measure what is easy to measure, not what is important; to emphasize effort rather than progress. As John Gall says in his book, **Systemantics**, given time, "Any system will produce reports indicating that it is meeting its objectives." This means that we tell people what they want to hear--that we are on time and under budget. Therefore, if your objective is to produce something on time and within budget, then that is probably what you will produce--"something" on time and within budget. If, however, your objective is to produce a system that solves the right problems, delivers the right results for the right operating costs within the minimum time, and is enthusiastically used by the client, then you have a more difficult set of problems that require radically different solutions.

Many people have tried to apply generally accepted project management techniques from other fields to systems development, without understanding what made project management in those fields different from that for systems development. As it turned out, the most important difference was the general lack of skills and standard procedures in systems development.

In most established design fields such as architecture or engineering, the individual practitioners are highly trained and are indoctrinated in standard practices on the job. In such an environment, one can apply project management to a well-understood process that will be implemented by trained professionals. In systems building, however, the situation is considerably different. Because of the relative newness of the field, the process is not well understood, nor are the practitioners well trained and experienced.

When we began to teach design principles, it became evident that we had to break down the process into small steps. At this level, structured systems development looks like most other methodologies (Fig. 7.2). The difference is the data that flows between phases (Fig. 7.3). Structured systems development aims at being "seamless," i.e., where the results of one phase lead naturally into the next; where data is not speci-

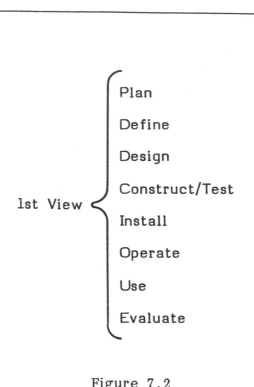

1st View
- Plan
- Define
- Design
- Construct/Test
- Install
- Operate
- Use
- Evaluate

Figure 7.2

fied redundantly (except for reliability); and where the end-product relates to the initial problem statement in a predictable and continuous manner.

A systems methodology is really a system to build systems. As such, it contains all the characteristics of any other system—outputs, processes, data bases, and inputs. Moreover, like any other system, a systems methodology must collect only data that will be used. Otherwise, the data will not be correct.

In **Structured Systems Development** I described the basic systems design procedure. At that point the design process focused on the output. But since then, by working backward, both figuratively and actually, we have developed the basic procedure we employ today in requirements definition.

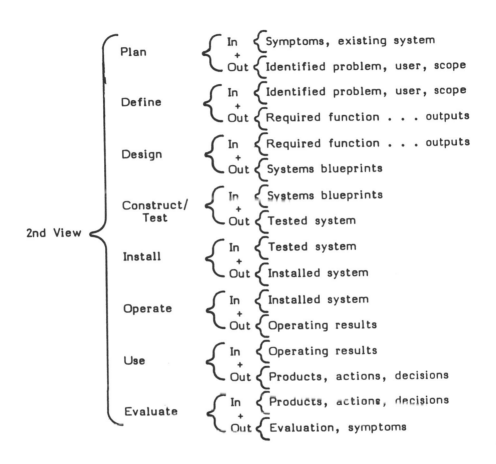

Figure 7.3

This book has been concerned with understanding requirements definition. Like any good structured procedure, requirements is bounded. It is bounded at the bottom by the systems design phase, and at the top by the systems planning phase. The beginning (or input) of the requirements definition phase is the end (or output) of the planning phase.

That beginning is variously called the "systems planning document" or "a defined application." We know, for example, that such a document must contain answers to at least the following questions:

* What are the real (as opposed to apparent) problems or opportunities the system is supposed to affect?
* Who are the users and what will the primary uses of the system be?
* What is the context and preliminary scope of the system?

Structured planning is now at the same developmental stage as requirements definition was five years ago and systems design five years before that. More and more organizations are recognizing the need to plan systems building carefully before launching into the development of major applications.

We keep pushing the process of logical, output-oriented, structured design further back. As we do, we begin to overlap with what has traditionally been management analysis. In the next decade, no doubt we will know even more. As someone put it, in ten years structured techniques will be taught in the board room.

Appendix/
Warnier/Orr Diagrams

****

Every statement about complexes
can be resolved into a statement about
their constituents and into the propositions
that describe the complexes completely.

Ludwig Wittgenstein
Tractatus Logico-Philosophicus

CHARACTERISTICS OF A GOOD DEFINITION LANGUAGE

During the last 25 years, several languages have been developed for the specification or definition of algorithmic or data structures. In recent years the "structured revolution" has produced several new forms of definition languages: HIPO charts, structure diagrams, Nassi-Shneiderman diagrams, SADT® diagrams, etc. Each has features to recommend it.

This appendix describes another tool for definition: the Warnier/Orr diagram. Warnier diagrams first appeared in **The Logical Construction of Programs** by Jean-Dominique Warnier. They were subsequently modified by myself (**Structured Systems Development**) and others to encompass the major features required of a definition language, resulting in Warnier/Orr diagrams.

This appendix delineates the characteristics desired in any good definition language, the syntax and semantics of Warnier/Orr diagrams, and the applications of Warnier/Orr diagrams to the definition of programs, systems, data structures, and data files.

In deciding upon a definition language, it is important to specify the principal characteristics that such a language should exhibit. An ideal language would be:

- Simple
- Logical
- Complete
- Graphic

All the elegant, and, therefore, most useful, definition languages are simple as well. BNF (Bachus-Naur Form), for example, has been used for over 20 years, mainly because of its elegance and simplicity.

A **simple** definition language contains a minimum of basic structured representations. Complex objects should be defined in terms of combinations and permutations of those basic representations.

Logical means that notation must be directly related to basic logical primitives. The basic representations of any definition language must be provable and, therefore, must tie back to the fundamental concepts of mathematical logic and set theory. One of Warnier's great contributions was to point out a manner in which this could be done systematically.

Complete means that the notation includes a sufficient set of structures to describe the vast majority of important structures and relationships.

Graphical means that a definition language will "picture" the solution as much as possible. Delineating the context as clearly as possible is a significant step in improving communications between user and vendor. Recent discoveries in the algebra of logical structures result from the ability to picture complex problems simply on a single sheet of paper. The way a problem is "seen" can have a major impact on the way the problem is perceived and, therefore, solved.

THE REQUIRED STRUCTURES

A great deal of importance has been placed on **sequence**, **alternation**, and **repetition** as basic structures, sufficient to define logical programs or data structures. This set of relations forms the basis for structured programming. Since these relations were first described, many additions to and modifications of the structures have been proposed. For the most part, the changes represent methods of reintroducing unconditional branches into program design. Our research indicates, however, that three additional logical relations must be represented if definitional language is to model the real world adequately: **hierarchy**, **concurrency**, and **recursion**.

Hierarchy (Invocation)

Of all the basic relations, hierarchy (invocation) is probably the most important. It is the one by which we can subdivide and conquer. Hierarchy is represented in Warnier/Orr diagrams by the use of the brace (curly bracket).

Program $\left\{\rule{0pt}{3em}\right.$

Figure 1. Hierarchy

From a static view, the bracket means "is defined as." From a dynamic view, the bracket shows the flow of control (in the case of processes), or the flow of access (in the case of data). In this respect, it is often valuable to show the upper and lower portions of the brace as two arrows indicating direction of flow.

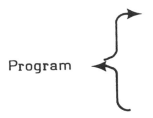

Figure 2. Flow of Control

Sequence

Sequence is shown by listing elements within a brace, one below another:

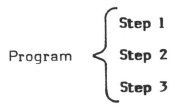

Figure 3. Sequence

Repetition

Repetition is represented by specifying, in paren-theses under the name of the element, the number of times that element is included in the definition.

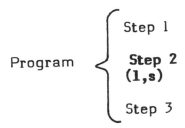

Figure 4. Repetition (DOUNTIL)

Theoretically, all elements have a number of times associated with them; however, in the normal case where the number of times is unity, this indication is ordinarily omitted.

Figure 4 shows repetition with at least one element, the "DOUNTIL" construct. In dealing with repetition, we want the capability to define a repetitive circumstance that may not contain any elements, the "DOWHILE" construct. This is done by making the first number within the number of times equal to zero.

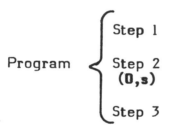

Program
{
Step 1

Step 2
(0,s)

Step 3
}

Figure 5.   Repetition (DOWHILE)

The degenerate case of repetitive definition occurs when an element may occur zero or one time, at most.

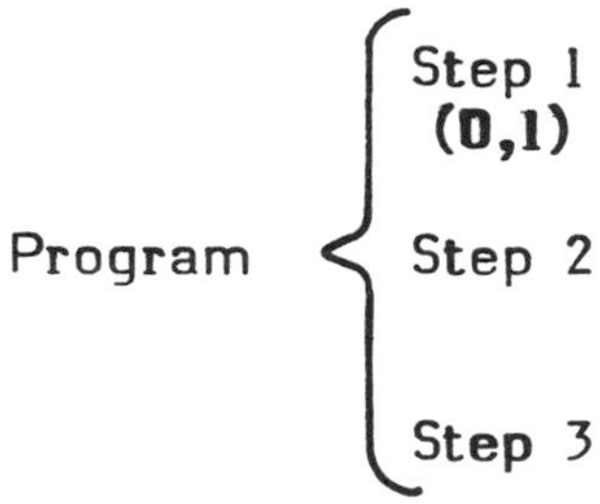

Program
{
Step 1
(0,1)

Step 2

Step 3
}

Figure 6.   Repetition (Optional)

Alternation

Alternation (selection) is represented by the use of the **exclusive or** operator: $\oplus$.

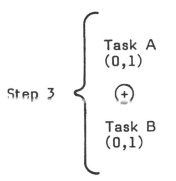

Figure 7. Alternation

Normally, the number of times is included in specifying alternative elements.

The negation sign, " —— ", is used over an element in alternation to show the negative case.

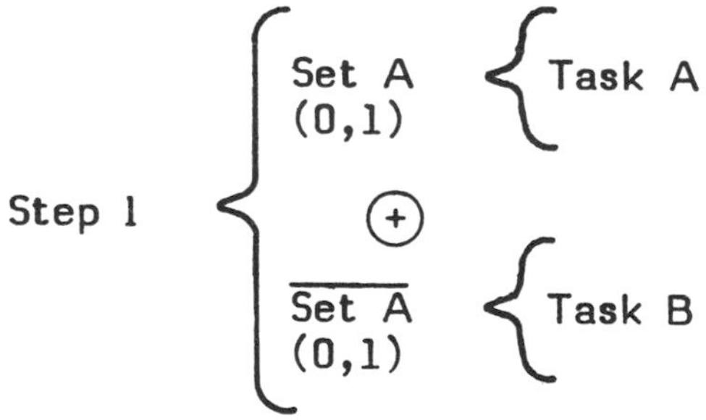

Figure 8. Alternation with Negation

A generalization of simple binary alternations is a definition for multiple alternation, "case" construct.

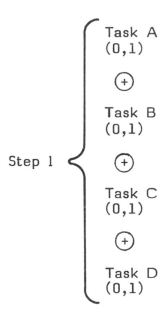

Figure 9.   Multiple Alternation (Case)

Concurrency

Concurrent definition is specified in Warnier/Orr diagrams by the use of the and/or operator "+". In concurrent structures, the elements are known, but the order is not.

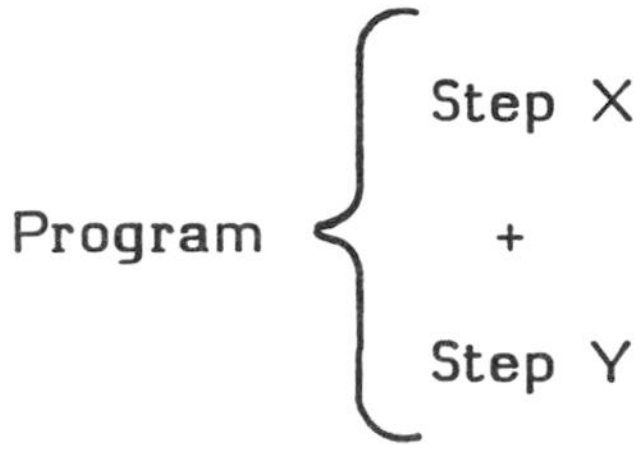

Program
{
   Step X

    +

   Step Y
}

Figure 10.   Concurrency

In Figure 10, for example, "Program" invokes both "Step X" and "Step Y," but the order of execution is independent. The definition implies, however, that both "Step X" and "Step Y" must be completed before "Program" is complete.

By using the number of times (0,1) with the and/or operator, it is possible to show a **logical or** structure.

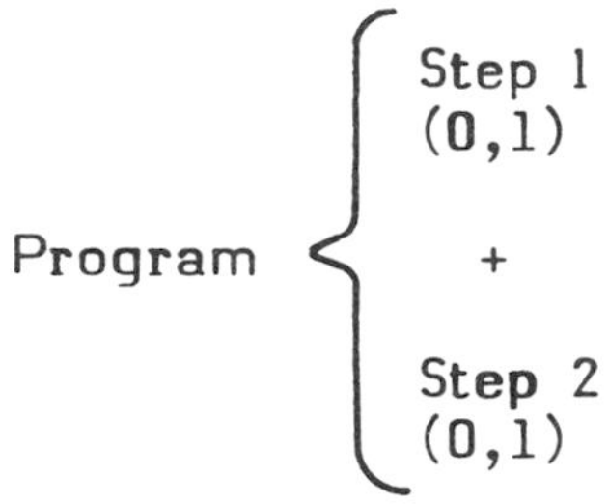

$$\text{Program} \left\{ \begin{array}{c} \text{Step 1} \\ (0,1) \\ \\ + \\ \\ \text{Step 2} \\ (0,1) \end{array} \right.$$

Figure 11. Logical Or

Recursion

Recursion is represented by using the name of an element in its own definition.

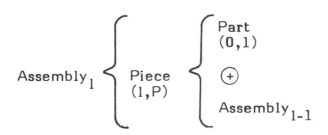

Figure 12. Recursion

The basic relations--hierarchy, sequence, repetition, alternation, concurrency, and recursion--form the foundation for using the Warnier/Orr diagram as a definitional tool.

SOME USEFUL EXTENSIONS TO THE BASIC LANGUAGE

A number of extensions to the logical structures have proved extremely useful. In particular, the symbology has been extended to include conventions for arithmetic operators, names, labels, and the use of operators with repetitive sets.

Arithmetic Operators

Extending the concept of concurrent elements has made it possible to include all the arithmetic operators. These operators are specified by the use of a box ☐ with the appropriate operator inside.

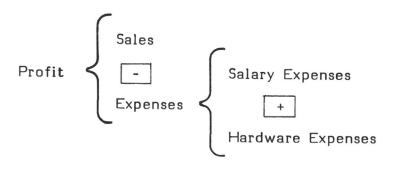

Figure 13

Names

Warnier/Orr diagrams use three types of names for elements: names of sets, of data elements, and of constants or literals. **Set names** appear with a brace to the right. **Data element names** appear with no brace to the right. **Literals** appear enclosed by quotation marks.

Type	Example
Set name	Customer
Data element	Customer.Name Customer.Address
Literal	"Total Sales"

Figure 14

Concatenated Names

In certain cases, it is useful to abbreviate a set or data element name by concatenating the name of the associated set. This is specified by the dot operator ".".

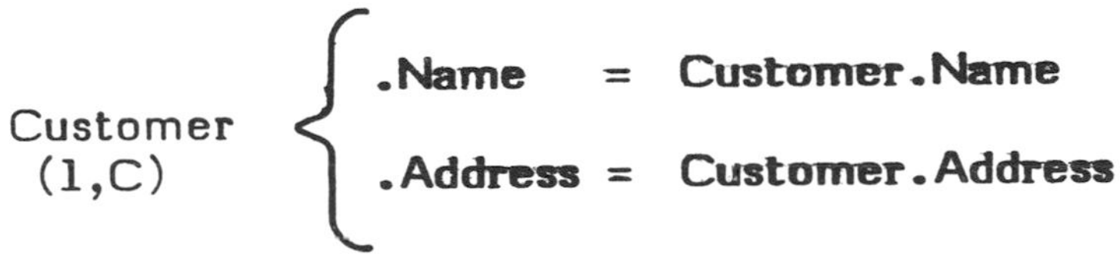

Customer
(1,C)
{
.Name = Customer.Name

.Address = Customer.Address

Figure 15. Concatenated Names

Labels

To distinguish clearly between the logical and physical names of routines, Warnier/Orr diagrams have been augmented by placing "bubbles" at the tops of braces. The bubbles contain the physical name (required by the operating system or compiler) of the process or data included.

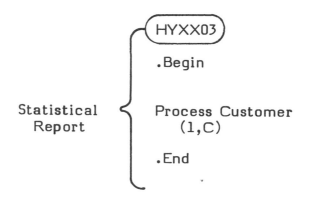

Figure 16.   Use of Physical Names

Keys

In specifying data structures, it is necessary to specify the ordering of the elements within a universal.   This is done by using an underline associated with an atomic element.   In these cases, the underline is placed under the appropriate atomic elements.   If there is more than one, it is assumed that they are sorted one within the other.   If this is not the case, then a number (subscript) is added after the underline to specify the order.   It is assumed that the sequence is ascending.   If this is not true, then the underline should be followed by a letter "D" before any number.

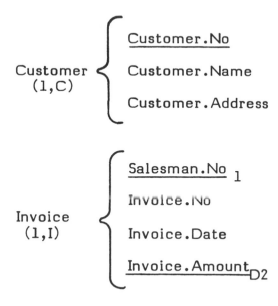

Figure 17

Operators Applied to Repetitive Elements

It is often valuable to be able to define an operation between a variable number of elements. This is done by placing the operator in front of a repetitive element.

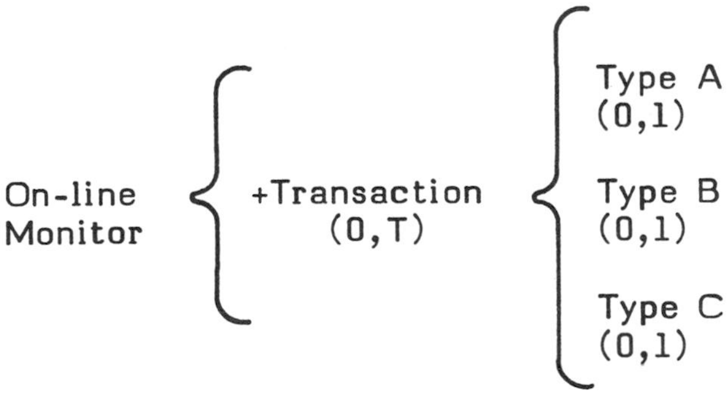

Figure 18

In Figure 18 "on-line monitor" is defined as a variable number of concurrent transactions, each of which is either "Type A," "Type B," or "Type C." Any operator, $\oplus$, $\boxplus$, +, etc., can be applied to a repetitive element.

Conditions

In moving from logical definitions to physical ones, it is necessary to be able not only to define the sets and subsets of data or processes to be dealt with, but also to deal with the conditions by which certain subsets could be specified. In order to do this, we have introduced a notation for specifying **conditions** within the diagrams.

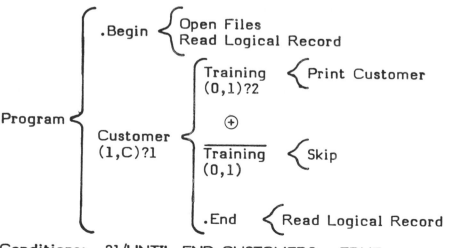

Conditions: ?1/UNTIL END-CUSTOMERS = TRUE
?2/IF CUSTOMER.TYPE = "T"

Figure 19. Conditions

Conditions are used to define conditional tests that specify the actual execution of either repetitive or alternative structures. The introduction of conditions has made it possible to use Warnier/Orr diagrams both for logical and physical design, and to wholly eliminate the need for pseudo code.

ON INTERPRETATION

Warnier/Orr diagrams can be used to represent both processes and data. In the first instance, they define the flow of **control** and in the latter, the flow of **access**.

In one sense, Warnier/Orr diagrams allow us to represent sets and hierarchies of sets. Take, for example, a simple hierarchical data structure of a report.

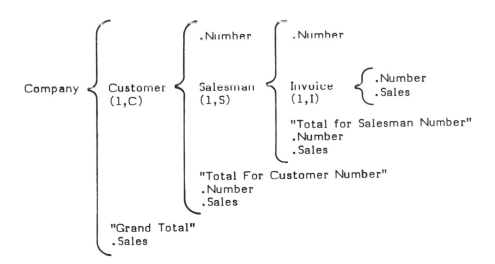

Figure 20.   Data Structure

The "data-structured" school of programming (Warnier, Jackson, Orr, Higgins) takes advantage of the isomorphism between data structure and process structure to develop program designs directly. For example, given a simple input structure (one record per invoice in sequence by customer, salesman, and invoice), the program structure (assuming at least one record on the input file) to produce the output in Figure 20 is identical to the data structure.

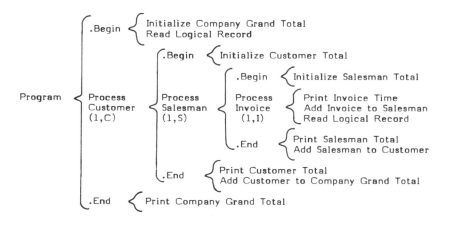

Figure 21

STRUCTURE CLASHES

The observation that process structure equals (or ought to equal) data structure has been noted many times in the history of programming. The utility of this fundamental concept seems to break down in many important cases, however.

For the most part, complexity in programs originates where the structure of the input and output are not identical. Jackson has characterized these discrepancies as structure clashes. In actuality, they are "hierarchy clashes."

Hierarchy clashes occur whenever two different, concurrent hierarchies are involved. For example, one of the most familiar hierarchy clashes is between the physical and logical hierarchies of a report.

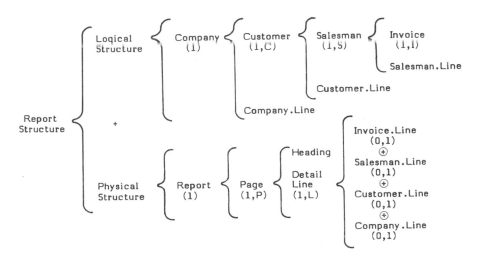

Figure 22

From the description in Figure 22, it is easy to see that two perfectly understandable and predictable hierarchies exist. The improved description of the problem makes it possible to form several general solutions, the most obvious one being the use of coroutines.

In a programming environment that does not have concurrent processing, a solution must be developed that fits one hierarchy within another through some logical or physical transformation. This must occur on either the logical hierarchy (Warnier, Orr) or on the code (Jackson). Warnier refers to this as "phase sets;" Orr and Jackson use the terms "inversion" or "flattening" to describe the transformation.

Structure clashes represent a major area of investigation. The ability to define concurrent structures, however, constitutes the most significant advance in being able to state the problem clearly.

The ability to show concurrence and recursion are by no means unique. Such relations could be added to any hierarchical diagramming technique. The fact that they were added to Warnier/Orr diagrams early allowed the solution of a series of difficult definitional problems. Moreover, the use of these concepts in conjunction with structures has made it possible to describe the logical structure of complex processes such as operating systems and teleprocessing systems.

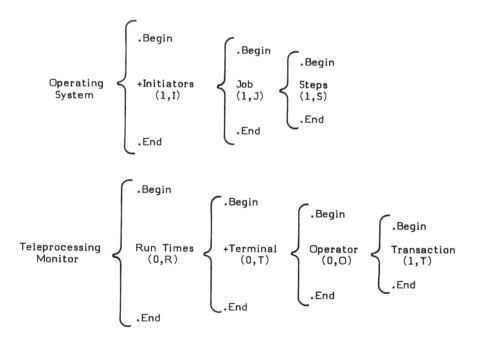

Figure 23

Although these systems demonstrate a great deal of overlap (apparent concurrency), there are only a few points of logical concurrency.

Jackson uses the concept of "state variable" to solve problems of concurrent structures. The traditional operating-system solution of saving and restoring program registers can be shown to be simply a means of implementing state-variables.

APPLICATIONS

Although Warnier/Orr diagrams were originally intended to define computer systems, they have been used for manual procedures as well. Because of their power and simplicity in communicating, the diagrams have proved to be most useful for this purpose.

Documenting Data Flow

There are two schools of structured design: (1) data structure and (2) data flow. Traditionally, hierarchy diagrams have proved weak in defining the flow of data. With Warnier/Orr diagrams, two alternative vehicles are available to represent data flow:

- In-out diagrams
- Assembly line diagrams

In-out diagrams are used to document data flow from an operational standpoint (Fig. 24).

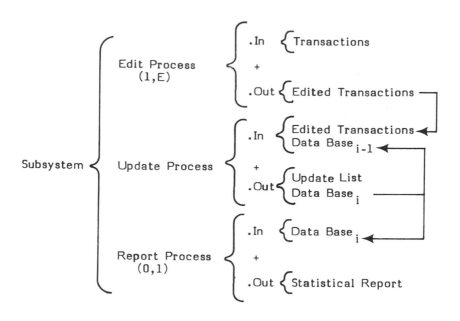

Figure 24. In-Out Diagram

Another method of representing data flow is the assembly line diagram. In this interpretation both data and process are shown to be concurrent and recursive. Each bracket on a diagram produces a single output.

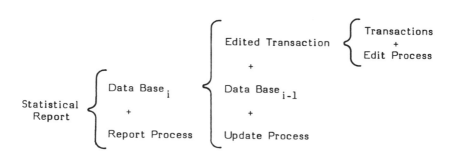

Figure 25

The Minimum, Complete Structures

There is always a temptation to add operators to definitional languages to deal with what appear to be unique circumstances. Unfortunately, this often leads to considerable difficulties in practice.

It is important to strive for only the minimum number of basic structures sufficient to represent the problem in a straightforward fashion. One can look at BNF and APL as opposite extremes in this respect.

BNF provides an illustration of exactly how far a minimum set of constructs can be carried when applied intelligently. But BNF also illustrates the problems that can occur when there are not enough basic structures. For example, because repetition was omitted from the basic structures of BNF, recursion had to be used, often inappropriately, to represent simple repetitive situations.

APL, on the other hand, shows both the power and the danger of introducing too many basic operations and combinations. The notation, though compact, is extremely complicated to understand.

Warnier/Orr diagrams have only a few more basic constructs than BNF and a fraction of those in APL. Warnier/Orr diagrams attack complexity through the use of hierarchy. By using a small number of precisely defined structures, the diagrams are capable of representing clearly a wide class of process and data structures.

Topology and Communication

How much information is appropriate to place on a diagram? Much has been made of the rule of "seven plus or minus two." Various individuals maintain this rule, obtained from psychology, puts a limit on the amount of information that can profitably be placed on a diagram.

Experience with very large Warnier/Orr diagrams tends to refute these arguments, however. Many users have produced wall-sized diagrams with hundreds of elements and dozens of levels. From the success of such diagrams, we can conclude that detail is not equivalent to complexity. The critical considerations for communication seem to be topology and context. Hierarchy diagrams are clearly superior to other forms of documentation, especially network forms, for this reason. As one proceeds down a hierarchy, it is possible to assume more about the current context. The mind needs only to deal with one thing at a time. By definition, tree diagrams have a clean topology, but perhaps their most important feature is the clarity of context.

Warnier/Orr Diagrams in Requirements Definition

The Warnier/Orr diagram is, next to the entity diagram, the principal tool used in requirements definition. The principal uses of the diagram are:

- to define functional data flows (assembly line diagram),
- to define process inputs-outputs (in-out diagram),
- to define decision processes, outputs, and cycles (standard diagram).

Warnier/Orr diagrams have proved to be a major new tool in the arsenal of the systems scientist.

Author's Note

The evolution of the Warnier/Orr diagram from a tool for analysis to a graphical specification language caught most of us involved unawares. Over the past few years, numerous additions have been made to the language to address specific uses, but these extensions to the syntax of the language have not been documented completely in any one place.

I decided to add this appendix for two reasons: (1) to provide a commonly available reference for analysts, programmers, managers, and users; and (2) to avoid cluttering up the main thread of the discussion of requirements definition.

Bibliography

****

Bohm, C., and Jacopini, G. "Flow diagrams, Turing machines, and languages with only two formation rules," **Communications of the ACM**, May 1966, 366-371.

Couger, J.D., and Knapp, R.W. **System analysis techniques**. New York: John Wiley & Sons, 1974.

Dijkstra, E.W. **A discipline of programming**. Englewood Cliffs, NJ: Prentice-Hall, 1976.

Doyle, P. **Every object is a system**. London: Interprint, 1976.

Freedman, D.P., and Weinberg, G.M. **Ethnotechnical review and techniques**. New York: Improved System Technologies, 1977.

Gall, J. **Systemantics**. New York: Simon & Schuster, 1978.

Hansen, P.B. **The architecture of concurrent programs**. Englewood Cliffs, NJ: Prentice-Hall, 1977.

Higgins, D.A. **Program design and construction**. Englewood Cliffs, NJ: Prentice-Hall, 1979.

Jackson, M.A. **Principles of program design**. New York: Academic Press, 1975.

Katzan, H. **Systems design and documentation: An introduction to the HIPO method**. New York: Van Nostrand Reinhold Co., 1976.

Kepner, C.H., and Tregoe, B.B. **The rational manager**. Princeton, NJ: Author, 1976.

Langer, S.K. **An introduction to symbolic logic**. New York: Dover Press, 1953.

McDaniel, H. **An introduction to decision logic tables**. Rev. ed. Princeton, NJ: Petrocelli Books, 1978.

Mills, H.D. "Principles of software engineering." **IBM Systems Journal**, December 1980, **19**(4), 414-420.

Myers, G.J. **Composite/structured design**. New York: Van Nostrand Reinhold Co., 1978.

Orr, K.T. **Structured systems development**. New York: Yourdon Press, 1977.

Ross, D.T. "Principles of structuring." **SofTech Technical Paper**, TP #082, November 1978.

--------. "Removing the limitations of natural language (with 'Principles behind the RSA language')." **Software Engineering**. New York: Academic Press, 1980.

Simon, H.A. **The science of the artificial**. Cambridge, MA: MIT Press, 1969.

Structured analysis and design. Infotech State of the Art Report, vol. 2, Invited Papers. Maidenhead, Eng.: Infotech International, 1978.

Sundgren, B. **Theory of data bases**. New York:
Petrocelli/Charter, 1975.

Warnier, J.-D. **Guide des utilisateurs du systeme
informatique**. Paris: Les editions d'organisation,
1979.

--------. **L'organisation des donnees d'un systeme**.
Paris: Les editions d'organisation, 1974.

--------. **Pratique de la construction d'un ensemble de
donnees**. Paris: Les editions d'organisation, 1976.

Wilson, M. "A semantics-based requirements and design
method." **IBM Technical Report**, September 1979.

Wittgenstein, L. **Tractatus Logico-Philosophicus**.
London: Routledge & Kegan Paul, 1961.

Yourdon, E., and Constantine, L.L. **Structured design**.
2nd ed. New York: Yourdon Press, 1978.

Index